WORKSHEETS
FOR CLASSROOM OR LAB PRACTICE

with contributions from

BEVERLY FUSFIELD

STEVE OUELLETTE

JAMES J. BALL
Indiana State University

BASIC COLLEGE MATHEMATICS
EIGHTH EDITION

Margaret L. Lial
American River College

Stanley A. Salzman
American River College

Diana L. Hestwood
Minneapolis Community and Technical College

Addison-Wesley
is an imprint of

Reproduced by Pearson Addison-Wesley from electronic files supplied by the author.

Publishing as Pearson Addison-Wesley, 75 Arlington Street, Boston, MA 02116.

ISBN-13: 978-0-321-57470-1
ISBN-10: 0-321-57470-2

2 3 4 5 6 BB 11 10 09

Addison-Wesley
is an imprint of

www.pearsonhighered.com

CONTENTS

Name: Date:
Instructor: Section:

Chapter 1 WHOLE NUMBERS

1.1 Reading and Writing Whole Numbers

Learning Objectives

1 Identify whole numbers.
2 Give the place value of a digit.
3 Write a number in words or digits.
4 Read a table.

Key Terms

Use the vocabulary terms listed below to complete each statement in exercises 1–3.

whole numbers **place value** **table**

1. A display of facts in rows and columns is called a table.

2. The whole numbers are 0, 1, 2, 3, 4, and so on.

3. The place value of each digit in a whole number is determined by its position in the whole number.

Objective 1 Identify whole numbers.

Indicate whether each number is a whole number or not a whole number.

1. 48 **1.** yes

2. 1.2 **2.** No

3. $6\frac{3}{4}$ **3.** No

4. 10,029 **4.** yes

Objective 2 Give the place value of a digit.

Write the digit for the given place value in each of the following whole numbers.

5. 9841 thousands **5.** 9

tens 1

6. 25,016 ten-thousands **6.** 2

hundreds 0

7. 186,321 hundred-thousands **7.** 1

ones 1

Name: Date:
Instructor: Section:

8.	5,813,207	millions	**8.**	5
		thousands		3
9.	2,800,439,012	billions	**9.**	2
		ten-millions		0

Write the digits for the given period (group) in each whole number

10.	29,176	thousands	**10.**	29
		ones		176
11.	75,229,301	millions	**11.**	75
		thousands		229
		ones		301
12.	70,000,603,214	billions	**12.**	70
		millions		000
		thousands		603
		ones		214
13.	300,459,200,005	billions	**13.**	300
		millions		459
		thousands		200
		ones		005

Objective 3 Write a number in words or digits

Write each number in words.

14. 8714 **14.** Eight thousand, Seven hundred Fourteen

15. 39,015 **15.** thirty nine thousand fifteen

16. 834,768 **16.** Eight hundred Thirty four thousand Seven hundred Sixty-Eight

Name: Date:
Instructor: Section:

17. 2,015,102

17. two million, fifteen thousand, one hundred two

18. 499,304,018

18. four hundred Ninety-nine million three hundred four thousand Eighteen

Write each number using digits.

19. Four thousand, one hundred twenty-seven

19. 4,127

20. Twenty-nine thousand, five hundred sixteen

20. 29,516

21. Six hundred eight-five million, two hundred fifty-nine

21. 685, 000, 259

22. Three hundred million, seventy-five thousand, two

22. 300, 075, 002

Write the numbers from each sentence using digits.

23. A bottle of a certain vaccine will give seven thousand, two hundred ten injections.

23. 7,210

24. Every year, nine hundred seventy-two thousand, four hundred thirty people visit a certain historical area.

24. 972,430

25. A supermarket has fifteen thousand three hundred thirteen different items for sale.

25. 15,313

26. The population of a large city is six million, two hundred five thousand.

26. 6,205,000

Name: Date:
Instructor: Section:

Objective 4 Read a table.

Use the table below for exercises 27–30. Write the number in digits.

Weight of Exerciser	123 lbs	130 lbs	143 lbs
	Calories burned in 30 minutes		
Cycling	168	177	195
Running	324	342	375
Jumping Rope	273	288	315
Walking	162	171	189

Source: Fitness magazine

27. The number of calories burned by a 130-pound adult in 30 minutes of cycling. **27.** 177

28. The activity which will burn at least 300 calories when performed by a 123-pound adult. **28.** running

29. The number of calories burned by a 143-pound adult in 30 minutes of jumping rope. **29.** 315

30. The weight of an adult who will burn 189 calories in 30 minutes of walking. **30.** 143 lbs

Name: Date:
Instructor: Section:

Chapter 1 WHOLE NUMBERS

1.2 Adding Whole Numbers

Learning Objectives

1. Add two single-digit numbers.
2. Add more than two numbers.
3. Add when regrouping (carrying) is not required.
4. Add with regrouping (carrying).
5. Use addition to solve application problems.
6. Check the answer in addition.

Key Terms

Use the vocabulary terms listed below to complete each statement in exercises 1–7.

addition **addends** **sum (total)** **commutative property of addition**

associative property of addition **regrouping** **perimeter**

1. By the________________________________, changing the order of the addends in an addition problem does not change the sum.

2. In addition, the numbers being added are called the ________________.

3. The process of finding the total is called __________________.

4. By the________________________________, changing the grouping of the addends in an addition problem does not change the sum.

5. If the sum of the digits in any column is greater than 9, use the process called ____________________.

6. The distance around the outside edges of a figure is called the ________________.

7. The answer to an addition problem is called the __________________.

Objective 1 Add two single-digit numbers.

Add.

1. $3 + 9$ 1. ________________

2. $8 + 7$ 2. ________________

Name: Date:
Instructor: Section:

Objective 2 Add more than two numbers.

Add.

3. 7
2
5
3
+ 8

3. ____________

4. 9
7
2
5
+ 6

4. ____________

5. 3
4
9
2
7
+ 6

5. ____________

6. 6
4
9
8
4
+ 3

6. ____________

Objective 3 Add when regrouping (carrying) is not required.

Add.

7. 42
+ 57

7. ____________

8. 421
+ 567

8. ____________

Name: Date:
Instructor: Section:

9. 86,305 + 12,672 **9.** ________

10. 45,158 20,340 + 2401 **10.** ________

Objective 4 Add with regrouping (carrying).

Add.

11. 83 + 29 **11.** ________

12. 563 + 478 **12.** ________

13. 7439 + 8376 **13.** ________

14. 7033 809 2532 + 41 **14.** ________

15. 3197 + 420 + 638 + 67 **15.** ________

16. 6835 + 97 + 246 + 4001 **16.** ________

Name: Date:
Instructor: Section:

Objective 5 Use addition to solve application problems.

Using the map below, find the shortest distance between the following cities. All distances are in miles.

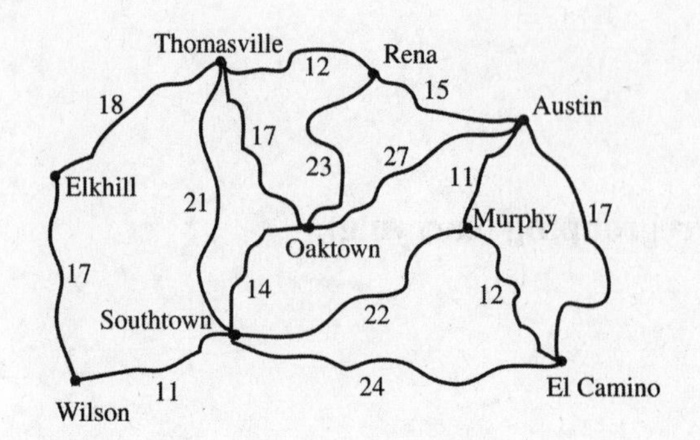

17. Murphy and Thomasville **17.** ______________

18. Wilson and Austin **18.** ______________

19. El Camino and Thomasville **19.** ______________

Solve the following application problems, using addition.

20. Kevin Levy has 52 nickels, 37 dimes, and 119 quarters. How many coins does he have altogether? **20.** ______________

21. The theater sold 276 adult tickets, and 349 child tickets. How many tickets were sold altogether? **21.** ______________

22. At a charity bazaar, a church has a total of 1873 books for sale, while a lodge has 3358 books for sale. How many books are for sale? **22.** ______________

Name: Date:
Instructor: Section:

Find the perimeter or total distance around each of the following figures.

23. **23.** ____________

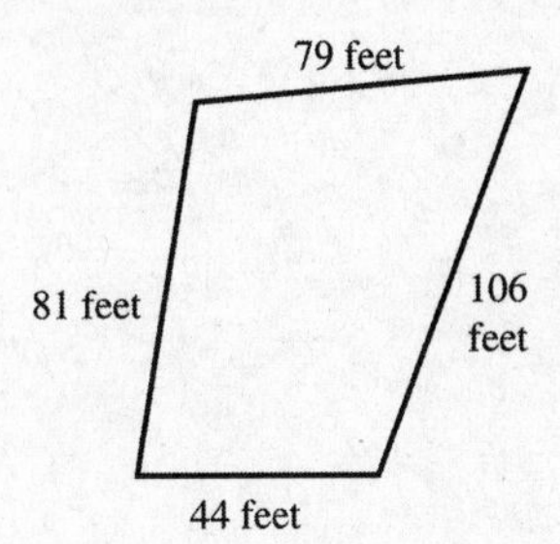

24. **24.** ____________

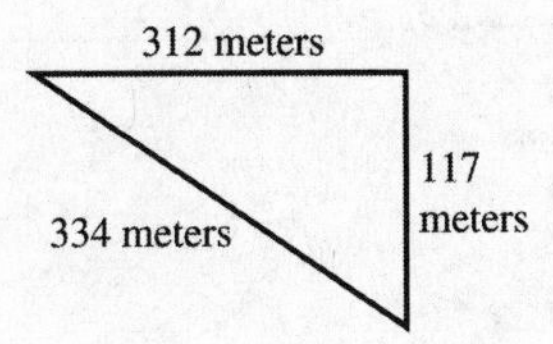

25. **25.** ____________

206 yards 197 yards
107 yards 107 yards
427 yards

Check the following additions. If an answer is incorrect, give the correct answer.

26. **26.** ____________

$$\begin{array}{r} 67 \\ 48 \\ +\ 83 \\ \hline 198 \end{array}$$

27. **27.** ____________

$$\begin{array}{r} 73 \\ 9815 \\ 390 \\ +\ 7002 \\ \hline 16,270 \end{array}$$

Name: Date:
Instructor: Section:

28.

$$\begin{array}{r} 723 \\ 681 \\ 29 \\ 412 \\ +\,103 \\ \hline 1947 \end{array}$$

28. ________________

29.

$$\begin{array}{r} 3028 \\ 335 \\ 2914 \\ 688 \\ +\,1647 \\ \hline 8612 \end{array}$$

29. ________________

30.

$$\begin{array}{r} 72 \\ 38 \\ 5735 \\ 764 \\ +\;\;16 \\ \hline 6625 \end{array}$$

30. ________________

Name: Date:
Instructor: Section:

Chapter 1 Whole Numbers

1.3 Subtracting Whole Numbers

Learning Objectives	
1	Change addition problems to subtraction and subtraction problems to addition.
2	Identify the minuend, subtrahend, and difference.
3	Subtract when no regrouping (borrowing) is needed.
4	Check subtraction answers by adding.
5	Subtract by regrouping (borrowing).
6	Solve application problems with subtraction.

Key Terms

Use the vocabulary terms listed below to complete each statement in exercises 1–4.

minuend **subtrahend** **difference** **regrouping**

1. The number from which another number is being subtracted is called the ____________.

2. In order to subtract 29 from 76, use a process called ____________ from the tens place.

3. The ____________ is the number being subtracted.

4. The answer to a subtraction problem is called the ____________.

Objective 1 Change addition problems to subtraction and subtraction problems to addition.

Write two subtraction problems for each addition problem.

1. 149 + 38 = 187 **1.** ____________

2. 478 + 239 = 717 **2.** ____________

Write an addition problem for each subtraction problem.

3. 1211 – 426 = 785 **3.** ____________

4. 5094 – 113 = 4981 **4.** ____________

Name: Date:
Instructor: Section:

Objective 2 Identify the minuend, subtrahend, and difference.

Identify the minuend, subtrahend, and difference in each of the following subtraction problems.

5. $98 - 36 = 62$

5.
Minuend ________
Subtrahend ________
Difference ________

6. $35 - 9 = 24$

6.
Minuend ________
Subtrahend ________
Difference ________

Objective 3 Subtract when no regrouping (borrowing) is needed.

Subtract.

7. $\begin{array}{r} 5573 \\ -\ 422 \\ \hline \end{array}$

7. ________

8. $\begin{array}{r} 8539 \\ -\ 2527 \\ \hline \end{array}$

8. ________

Objective 4 Check subtraction answers by adding.

Check the following subtractions. If an answer is not correct, give the correct answer.

9. $\begin{array}{r} 192 \\ -\ 39 \\ \hline 167 \end{array}$

9. ________

10. $\begin{array}{r} 4847 \\ -\ 3768 \\ \hline 1121 \end{array}$

10. ________

11. $\begin{array}{r} 5763 \\ -\ 2783 \\ \hline 3980 \end{array}$

11. ________

Name: Date:
Instructor: Section:

12. 31,146
− 7312
23,834

12. ________________

13. 82,004
− 3917
79,193

13. ________________

Objective 5 Subtract by regrouping (borrowing).

Subtract.

14. 927
− 729

14. ________________

15. 613
− 421

15. ________________

16. 4687
− 2769

16. ________________

17. 33,728
− 7829

17. ________________

18. 86,372
− 29,485

18. ________________

19. 302
− 57

19. ________________

20. 7000
− 297

20. ________________

Name: Date:
Instructor: Section:

Objective 6 Solve application problems with subtraction.

Solve each application problem.

21. A Girl Scout has 52 boxes of cookies to sell. If she sells 27 boxes, how many boxes will she have left?

21. ______________

22. An airplane is carrying 234 passengers. When it lands in Atlanta, 139 passengers get off the plane. How many passengers are then left on the plane?

22. ______________

23. Nathaniel Best has $553 in his checking account. He writes a check for $134. How much is then left in the account?

23. ______________

24. On Sunday, 7342 people went to a football game, while on Monday, 9138 people went. How many more people went on Monday?

24. ______________

25. Sally Tanner had $22,143 withheld from her paycheck last year for income tax. She actually owes only $16,959 in tax. What refund should she receive?

25. ______________

26. The Conrads now pay $439 per month for rent. If they rent a larger apartment, the payment will be $702 per month. How much extra will they pay each month?

26. ______________

Name: Date:
Instructor: Section:

27. On Friday, 11,594 people visited Eastridge Amusement Park, while 14,352 people visited the park on Saturday. How many more people visited the park on Saturday?

27. ________________

28. One bid for painting a house was $2134. A second bid was $1954. How much would be saved using the second bid?

28. ________________

29. Last year, 574 athletes competed in a district tract meet at Johnson College. This year, 498 athletes competed. How many fewer athletes competed this year than last?

29. ________________

30. At People's Bank, Marc Lukas can earn $1538 per year in interest, while Farmer's Bank would pay him $1643 interest. How much additional interest would he earn at the second bank?

30. ________________

Name: Date:
Instructor: Section:

Chapter 1 WHOLE NUMBERS

1.4 Multiplying Whole Numbers

Learning Objectives

1. Identify the parts of a multiplication problem.
2. Do chain multiplications.
3. Multiply by single-digit numbers.
4. Use multiplication shortcuts for numbers ending in zeros.
5. Multiply by numbers having more than one digit.
6. Solve application problems with multiplication.

Key Terms

Use the vocabulary terms listed below to complete each statement in exercises 1–6.

factors **product** **commutative property of multiplication**

associative property of multiplication **chain multiplication problem**

multiple

1. By the________________________________, changing the order of the factors in a multiplication problem does not change the product.

2. In multiplication, the numbers being multiplied are called the ________________.

3. By the________________________________, changing the grouping of the factors in a multiplication problem does not change the product.

4. The product of two whole number factors is called a ______________ of either factor.

5. The answer to a multiplication problem is called the ________________.

6. A multiplication problem with more than two factors is a ________________________________.

Objective 1 Identify the parts of a multiplication problem.

Identify the factors and the product in each multiplication problem.

1. $5(2) = 10$

1\.
Factors ______________

Product ______________

Name: Date:
Instructor: Section:

2. $108 = 9 \times 12$

2.
Factors ______________

Product ______________

Objective 2 Do chain multiplications.

Multiply.

3. $4 \times 4 \times 2$ **3.** ______________

4. $3 \times 4 \times 7$ **4.** ______________

5. $(6)(4)(8)$ **5.** ______________

Objective 3 Multiply by single-digit numbers.

Multiply.

6. $\begin{array}{r} 54 \\ \times\ 4 \\ \hline \end{array}$ **6.** ______________

7. $\begin{array}{r} 163 \\ \times\ 5 \\ \hline \end{array}$ **7.** ______________

8. $\begin{array}{r} 405 \\ \times\ 7 \\ \hline \end{array}$ **8.** ______________

9. $\begin{array}{r} 31{,}763 \\ \times\ \quad 9 \\ \hline \end{array}$ **9.** ______________

10. $\begin{array}{r} 30{,}009 \\ \times\ \quad 6 \\ \hline \end{array}$ **10.** ______________

Name: Date:
Instructor: Section:

Objective 4 Use multiplication shortcuts for numbers ending in 0s.

Multiply.

11. 439×1000 **11.** ________________

12. $(852)(30)$ **12.** ________________

13. 3005×2000 **13.** ________________

14. 500×40 **14.** ________________

15. 8234×2000 **15.** ________________

Objective 5 Multiply by numbers having more than one digit.

Multiply.

16. $\begin{array}{r} 644 \\ \times\ 19 \\ \hline \end{array}$ **16.** ________________

17. $\begin{array}{r} 4031 \\ \times\ \ 48 \\ \hline \end{array}$ **17.** ________________

Name: Date:
Instructor: Section:

18. $\begin{array}{r} 7165 \\ \times\ 53 \\ \hline \end{array}$ **18.** ______________

19. $\begin{array}{r} 5249 \\ \times\ 63 \\ \hline \end{array}$ **19.** ______________

20. $\begin{array}{r} 8621 \\ \times\ 131 \\ \hline \end{array}$ **20.** ______________

Objective 6 Solve application problems with multiplication.

Solve the following application problems.

21. A fabric store has 16 bolts of silk. Each bolt contains 35 yards of silk. How many yards of silk does the fabric store have in all? **21.** ______________

22. On a recent trip the Jensen family drove 45 miles per hour on the average. They drove 22 hours altogether. How many miles did they drive altogether? **22.** ______________

23. Marisa Taylor saves $38 out of every pay check. Last year she received 24 pay checks. How much did she save? **23.** ______________

Name: Date:
Instructor: Section:

24. Heinen's Supermarket received a shipment of 28 cartons of canned vegetables. There were 24 cans in each carton. How many cans were there altogether?

24. ________________

25. The 2008 Toyota Prius is estimated to get 48 miles per gallon in city driving. Its fuel tank holds approximately 12 gallons of gasoline. How far can it travel on one tank of gasoline?

25. ________________

Find the total cost of the following items.

26. 18 chairs at $42 per chair

26. ________________

27. 512 boxes of chalk at $19 per box

27. ________________

28. 47 watches at $29 per watch

28. ________________

29. 178 baseball caps at $9 per cap

29. ________________

30. 79 clocks at $198 per clock

30. ________________

Name: Date:
Instructor: Section:

Chapter 1 WHOLE NUMBERS

1.5 Dividing Whole Numbers

Learning Objectives

1	Write division problems in three ways.
2	Identify the parts of a division problem.
3	Divide 0 by a number.
4	Recognize that a number cannot be divided by 0.
5	Divide a number by itself.
6	Divide a number by 1.
7	Use short division.
8	Use multiplication to check the answer to a division problem.
9	Use tests for divisibility.

Key Terms

Use the vocabulary terms listed below to complete each statement in exercises 1–5.

dividend **divisor** **quotient** **short division** **remainder**

1. The number left over when two numbers do not divide exactly is the ________________.

2. The number being divided by another number in a division problem is the ____________________.

3. The answer to a division problem is called the ________________.

4. In the problem $639 \div 9$, 9 is called the ________________.

5. ____________________ is a method of dividing a number by a one-digit divisor.

Objective 1 Write division problems in three ways.

Write each division problem using two other symbols.

1. $15 \div 3 = 5$ 1. ________________

2. $\frac{50}{25} = 2$ 2. ________________

Name: Date:
Instructor: Section:

Objective 2 Identify the parts of a division problem.

Identify the dividend, divisor, and quotient.

3. $63 \div 7 = 9$

3.
Dividend ____________
Divisor ____________
Quotient ____________

4. $5\overline{)30}$ with quotient 6

4.
Dividend ____________
Divisor ____________
Quotient ____________

5. $\frac{44}{11} = 4$

5.
Dividend ____________
Divisor ____________
Quotient ____________

Objective 3 Divide 0 by a number
Objective 4 Recognize that a number cannot be divided by 0.

Divide. If the division is not possible, write "**undefined.**"

6. $12\overline{)0}$ **6.** ____________

7. $0\overline{)72}$ **7.** ____________

8. $\frac{0}{6}$ **8.** ____________

9. $\frac{7}{0}$ **9.** ____________

10. $0 \div 15$ **10.** ____________

11. $9 \div 0$ **11.** ____________

Name: Date:
Instructor: Section:

Objective 5 Divide a number by itself
Objective 6 Divide a number by 1.

Divide.

12. $18 \div 18$ **12.** ______________

13. $1\overline{)38}$ **13.** ______________

Objective 7 Use short division.

Divide by using short division.

14. $2\overline{)84}$ **14.** ______________

15. $724 \div 5$ **15.** ______________

16. $\frac{651}{9}$ **16.** ______________

17. $8\overline{)1135}$ **17.** ______________

18. $984 \div 6$ **18.** ______________

19. $\frac{512}{3}$ **19.** ______________

Objective 8 Use multiplication to check the answer to a division problem.

Use multiplication to check each answer. If an answer is incorrect, find the correct answer.

20. $\begin{array}{r} 1522 \text{ R4} \\ 6\overline{)9137} \end{array}$ **20.** ______________

21. $3852 \div 4 = 963$ **21.** ______________

22. $\frac{8621}{3} = 2873 \text{ R } 2$ **22.** ______________

23. $\begin{array}{r} 4814 \\ 7\overline{)40{,}698} \end{array}$ **23.** ______________

Name: Date:
Instructor: Section:

24. $\dfrac{18{,}150}{3} = 650$ **24.** ______________

25. $20{,}351 \div 6 = 3391$ R 5 **25.** ______________

Objective 9 Use tests for divisibility.

Determine if the following numbers are divisible by 2, 3, 5, or 10 Write **yes** *or* **no***.*

26. 50

26. **2:**______________
3:______________
5:______________
10:______________

27. 897

27. **2:**______________
3:______________
5:______________
10:______________

28. 908

28. **2:**______________
3:______________
5:______________
10:______________

29. 6205

29. **2:**______________
3:______________
5:______________
10:______________

30. 32,175

30. **2:**______________
3:______________
5:______________
10:______________

Name: Date:
Instructor: Section:

Chapter 1 WHOLE NUMBERS

1.6 Long Division

Learning Objectives

1. Do long division.
2. Divide numbers ending in 0 by numbers ending in 0.
3. Use multiplication to check division answers.

Key Terms

Use the vocabulary terms listed below to complete each statement in exercises 1–5.

long division **dividend** **divisor** **quotient** **remainder**

1. In the problem $751 \div 23 = 32 \text{ R } 15$, 751 is called the ____________________.

2. In the problem $751 \div 23 = 32 \text{ R } 15$, 15 is called the ____________________.

3. In the problem $751 \div 23 = 32 \text{ R } 15$, 23 is called the ____________________.

4. In the problem $751 \div 23 = 32 \text{ R } 15$, 32 is called the ____________________.

5. ____________________ is a method of dividing a number by a divisor with more than one digit.

Objective 1 Do long division.

Divide using long division. Check each answer.

1. $32\overline{)2624}$ **1.** ________________

2. $29\overline{)9396}$ **2.** ________________

3. $42\overline{)3234}$ **3.** ________________

Name: Date:
Instructor: Section:

4. $23\overline{)1587}$ **4.** ________________

5. $53\overline{)5406}$ **5.** ________________

6. $37\overline{)4215}$ **6.** ________________

7. $89\overline{)7649}$ **7.** ________________

8. $56\overline{)9314}$ **8.** ________________

9. $94\overline{)29,047}$ **9.** ________________

10. $71\overline{)412,794}$ **10.** ________________

Name: Date:
Instructor: Section:

11. $28\overline{)177{,}919}$ **11.** ________________

12. $86\overline{)8{,}473{,}758}$ **12.** ________________

13. $205\overline{)6{,}680{,}335}$ **13.** ________________

14. $327\overline{)98{,}413{,}712}$ **14.** ________________

15. $657\overline{)429{,}700}$ **15.** ________________

16. $732\overline{)4{,}268{,}292}$ **16.** ________________

Name: Date:
Instructor: Section:

Objective 2 Divide numbers ending in 0 by numbers ending in 0.

Divide.

17. $80\overline{)560}$ **17.** ____________

18. $400\overline{)6000}$ **18.** ____________

19. $2000\overline{)12{,}000}$ **19.** ____________

20. $800\overline{)10{,}400}$ **20.** ____________

21. $910\overline{)38{,}220}$ **21.** ____________

22. $750\overline{)25{,}500}$ **22.** ____________

23. $1200\overline{)960{,}000}$ **23.** ____________

Objective 3 Use multiplication to check division answers.

Check each answer. If an answer is incorrect, give the correct answer.

24. $\begin{array}{r} 87 \text{ R}16 \\ 37\overline{)3235} \end{array}$ **24.** ____________

25. $\begin{array}{r} 65 \text{ R}5 \\ 89\overline{)5790} \end{array}$ **25.** ____________

Name: Date:
Instructor: Section:

26. $\begin{array}{r}346 \text{ R}18\\ 74\overline{)25{,}621}\end{array}$

26. ________________

27. $\begin{array}{r}44 \text{ R}22\\ 103\overline{)4658}\end{array}$

27. ________________

28. $\begin{array}{r}231 \text{ R}183\\ 205\overline{)47{,}538}\end{array}$

28. ________________

29. $\begin{array}{r}297 \text{ R}79\\ 318\overline{)94{,}207}\end{array}$

29. ________________

30. $\begin{array}{r}400 \text{ R}30\\ 428\overline{)196{,}883}\end{array}$

30. ________________

Name: Date:
Instructor: Section:

Chapter 1 WHOLE NUMBERS

1.7 Rounding Whole Numbers

Learning Objectives

1. Locate the place to which a number is to be rounded.
2. Round numbers.
3. Round numbers to estimate an answer.
4. Use front end rounding to estimate an answer.

Key Terms

Use the vocabulary terms listed below to complete each statement in exercises 1–3.

rounding **estimate** **front end rounding**

1. ________________________ is rounding to the highest possible place so that all the digits become zeros except the first one.

2. In order to find a number that is close to the original number, but easier to work with, use a process called ____________________.

3. ____________________ to find an answer close to the exact answer.

Objective 1 Locate the place to which a number is to be rounded.

Locate the place to which the number is rounded by underlining the appropriate digit.

1. 257,301 Nearest ten **1.** ________________

2. 1037 Nearest hundred **2.** ________________

3. 645,371 Nearest ten-thousand **3.** ________________

4. 39,943,712 Nearest million **4.** ________________

Objective 2 Round numbers.

Round each number as indicated.

5. 7863 to the nearest hundred **5.** ________________

6. 1382 to the nearest ten **6.** ________________

7. 18,211 to the nearest hundred **7.** ________________

8. 9348 to the nearest hundred **8.** ________________

9. 8398 to the nearest hundred **9.** ________________

10. 41,099 to the nearest hundred **10.** ________________

Name: Date:
Instructor: Section:

11. 51,803 to the nearest thousand **11.** ______________

12. 16,968 to the nearest hundred **12.** ______________

13. 53,595 to the nearest hundred **13.** ______________

14. 476,943 to the nearest ten-thousand **14.** ______________

15. 576,295 to the nearest hundred-thousand **15.** ______________

16. 14,823,307 to the nearest million **16.** ______________

Objective 3 Round numbers to estimate an answer.

Estimate each answer by rounding to the nearest ten. Then find the exact answer.

17.
$$\begin{array}{r} 37 \\ 24 \\ 58 \\ +\ 91 \\ \hline \end{array}$$

17.
Estimate ______________
Exact ______________

18.
$$\begin{array}{r} 19 \\ 87 \\ 35 \\ +\ 20 \\ \hline \end{array}$$

18.
Estimate ______________
Exact ______________

19.
$$\begin{array}{r} 69 \\ -\ 42 \\ \hline \end{array}$$

19.
Estimate ______________
Exact ______________

20.
$$\begin{array}{r} 88 \\ -\ 52 \\ \hline \end{array}$$

20.
Estimate ______________
Exact ______________

Name: Date:
Instructor: Section:

Estimate each answer by rounding to the nearest hundred. Then find the exact answer.

21.
$$\begin{array}{r} 276 \\ 312 \\ 174 \\ +\ 936 \\ \hline \end{array}$$

21.
Estimate ______________

Exact ______________

22.
$$\begin{array}{r} 419 \\ 188 \\ 324 \\ +\ 194 \\ \hline \end{array}$$

22.
Estimate ______________

Exact ______________

23.
$$\begin{array}{r} 971 \\ -\ 382 \\ \hline \end{array}$$

23.
Estimate ______________

Exact ______________

24.
$$\begin{array}{r} 815 \\ -\ 678 \\ \hline \end{array}$$

24.
Estimate ______________

Exact ______________

25.
$$\begin{array}{r} 912 \\ \times\ 784 \\ \hline \end{array}$$

25.
Estimate ______________

Exact ______________

26.
$$\begin{array}{r} 876 \\ \times\ 141 \\ \hline \end{array}$$

26.
Estimate ______________

Exact ______________

Name: Date:
Instructor: Section:

Objective 4 Use front end rounding to estimate an answer.

Estimate each answer using front end rounding. Then find the exact answer.

27.
$$\begin{array}{r} 571 \\ 42 \\ 215 \\ +\ 2452 \\ \hline \end{array}$$

27.
Estimate ____________

Exact ____________

28.
$$\begin{array}{r} 313 \\ -\ 49 \\ \hline \end{array}$$

28.
Estimate ____________

Exact ____________

29.
$$\begin{array}{r} 980 \\ \times\ 37 \\ \hline \end{array}$$

29.
Estimate ____________

Exact ____________

30.
$$\begin{array}{r} 437 \\ \times\ 29 \\ \hline \end{array}$$

30.
Estimate ____________

Exact ____________

Name: Date:
Instructor: Section:

Chapter 1 WHOLE NUMBERS

1.8 Exponents, Roots, and Order of Operations

Learning Objectives
1 Identify an exponent and a base.
2 Find the square root of a number.
3 Use the order of operations.

Key Terms

Use the vocabulary terms listed below to complete each statement in exercises 1–3.

square root **perfect square** **order of operations**

1. For problems or expressions with more than one operation, the ______________________ tells what to do first, second, and so on, to obtain the correct answer.

2. The ______________________ of a whole number is the number that can be multiplied by itself to produce the given number.

3. A ______________________ is a number that is the square of a whole number.

Objective 1 Identify an exponent and a base.

Identify the exponent and the base, then simplify each expression.

1. 7^2

1. Exponent ________

Base ________

Expression ________

2. 2^7

2. Exponent ________

Base ________

Expression ________

3. 8^3

3. Exponent ________

Base ________

Expression ________

Name: Date:
Instructor: Section:

4. 10^4

4. Exponent ________

Base ________

Expression ________

Objective 2 Find the square root of a number.

Find each square root.

5. $\sqrt{16}$ **5.** ________

6. $\sqrt{64}$ **6.** ________

7. $\sqrt{121}$ **7.** ________

8. $\sqrt{169}$ **8.** ________

9. $\sqrt{225}$ **9.** ________

Fill in each blank.

10. $18^2 =$ _____ so $\sqrt{\_\_\_\_\_} = 18$ **10.** ________

11. $50^2 =$ _____ so $\sqrt{\_\_\_\_\_} = 50$ **11.** ________

12. $25^2 =$ _____ so $\sqrt{\_\_\_\_\_} = 25$ **12.** ________

13. $20^2 =$ _____ so $\sqrt{400} =$ _____ **13.** ________

14. $36^2 =$ _____ so $\sqrt{1296} =$ _____ **14.** ________

Objective 3 Use the order of operations.

Simplify each expression using the order of operations.

15. $6^2 + 5 - 2$ **15.** ________

16. $2^4 + 3 \cdot 4 - 5$ **16.** ________

Name: Date:
Instructor: Section:

17. $6 \cdot 5 - 5 \div 0$

17. ______________

18. $8 \cdot 9 \div 12$

18. ______________

19. $9 \cdot 7 - 3 \cdot 12$

19. ______________

20. $7 - 25 \div 5$

20. ______________

21. $6 \cdot 3^2 + 0 \div 6$

21. ______________

22. $8 \cdot 5 - 12 \div (2 \cdot 3 - 6)$

22. ______________

23. $4 \cdot 3 + 8 \cdot 5 - 7$

23. ______________

24. $4 \cdot (9 - 7) + 3 \cdot 8$

24. ______________

25. $2^3 \cdot 3^2 + 5(3) \div 5$

25. ______________

26. $6 \cdot \sqrt{144} - 6 \cdot 8$

26. ______________

Name: Date:
Instructor: Section:

27. $42 \div 6 + 3 \cdot \sqrt{49}$

27. ________________

28. $2 \cdot \sqrt{121} - 2 \div \sqrt{4} + (14 - 2 \cdot 7) \div 4$

28. ________________

29. $25 \div 5 \cdot 3 \cdot 9 \div (14 - 11)$

29. ________________

30. $3^2 \cdot \sqrt{36} \div \sqrt{81} \div 3 + 2 \cdot 3 - 2$

30. ________________

Name: Date:
Instructor: Section:

Chapter 1 WHOLE NUMBERS

1.9 Reading Pictographs, Bar Graphs, and Line Graphs

Learning Objectives
1. Read and understand a pictograph.
2. Read and understand a bar graph.
3. Read and understand a line graph.

Key Terms

Use the vocabulary terms listed below to complete each statement in exercises 1–3.

pictograph **bar graph** **line graph**

1. A ______________________ is used to display a trend.

2. A graph that uses pictures or symbols to display information is called a ______________________.

3. A ______________________ uses bars of various heights to show quantity.

Objective 1 Read and understand a pictograph.

The pictograph shows the amount of sales tax in various states. Use the pictograph to answer exercises 1–5.

State Sales Tax

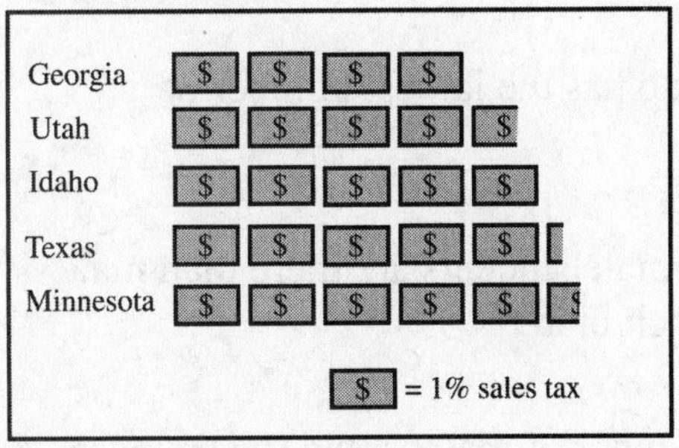

Georgia
Utah
Idaho
Texas
Minnesota

$ = 1% sales tax

Source: Federation of Tax Administrators

1. Which state shown in the pictograph charges the least sales tax? 1. ______________

2. Which state has a sales tax of 5%? 2. ______________

3. According to the pictograph, which state has the greatest sales tax? 3. ______________

Name: Date:
Instructor: Section:

4. Which state has a sales tax of 4%? **4.** ____________

5. By about how much does Minnesota's sales tax exceed Utah's sales tax? **5.** ____________

The pictograph shows the number of male and female students in the school's language clubs. Use the pictograph to answer exercises 6–10.

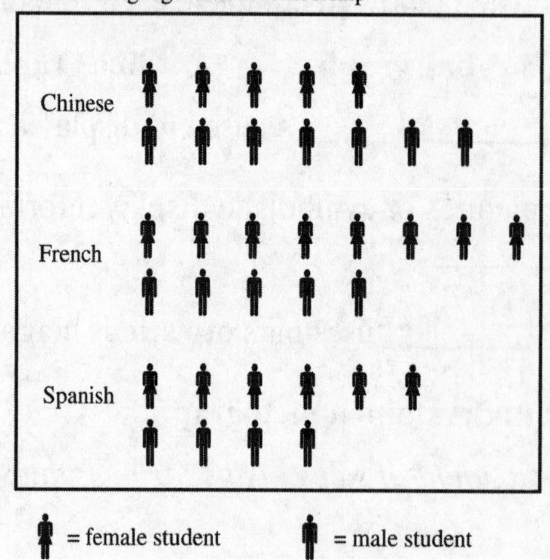

6. Which language club has the largest number of members? **6.** ____________

7. How many more female students are there than male students in the French club? **7.** ____________

8. What is the total number of members in the Chinese club? **8.** ____________

9. Which language club has the least number of members? **9.** ____________

10. How many more female students are in the French and Spanish clubs combined than male students? **10.** ____________

Name: Date:
Instructor: Section:

Objective 2 Read and understand a bar graph.

The bar graph shows the number of ice cream cones of different flavors that were sold at a barbecue. Use the bar graph to answer exercises 11–15.

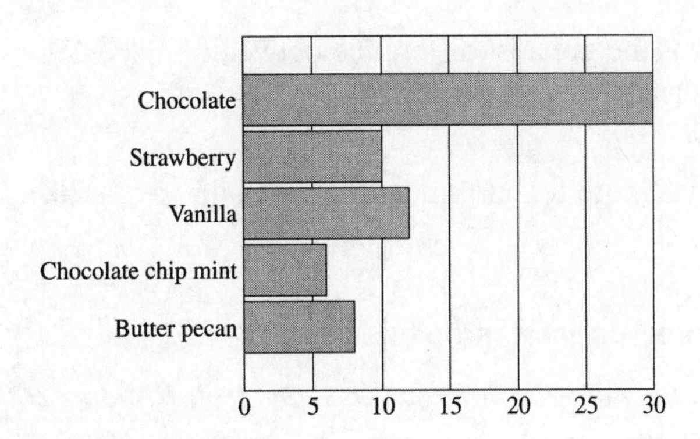

11. How many strawberry cones were sold? **11.** ______________

12. Which flavor had the fewest sales? **12.** ______________

13. How many more vanilla cones were sold than butter pecan cones? **13.** ______________

14. How many more chocolate cones were sold than vanilla and strawberry combined? **14.** ______________

15. How many strawberry and butter pecan cones were sold in total? **15.** ______________

The bar graph shows the enrollment by gender in each class at a small college. Use the bar graph to answer questions 16–20.

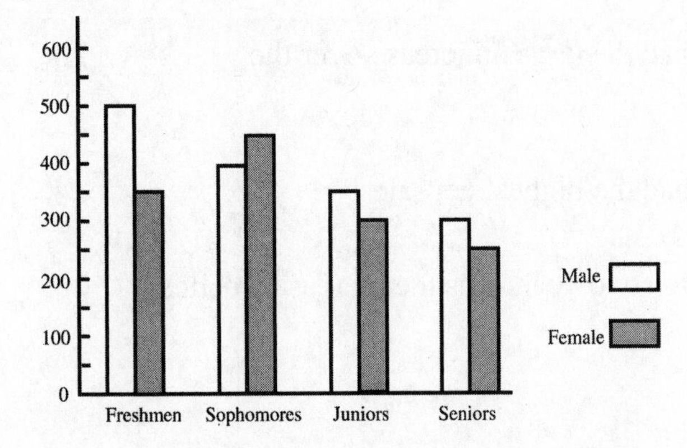

16. How many more male freshmen are there than female seniors? **16.** ______________

Name: Date:
Instructor: Section:

17. Find the total number of students enrolled. **17.** ____________

18. How many more sophomores are there than juniors? **18.** ____________

19. Which class has the greatest difference between male students and female students? **19.** ____________

20. Which class has more female students than male students? **20.** ____________

Objective 3 Read and understand a line graph.

The line graph shows the net sales for Ajax Systems from 2003 to 2007. Use the line graph to answer exercises 21–25.

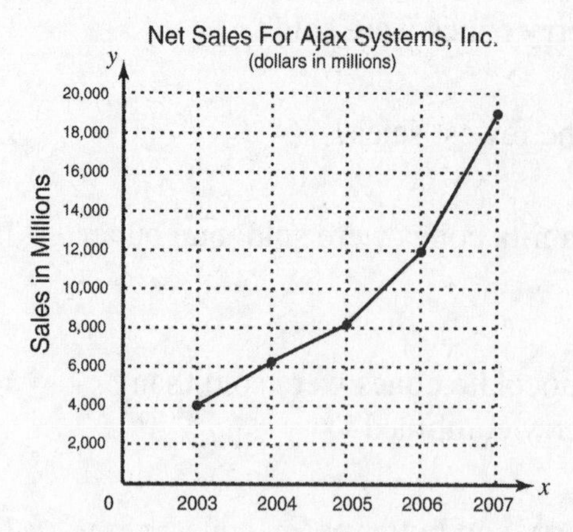

21. What trend or pattern is shown in the graph? **21.** ____________

22. Approximately what were the net sales in 2003? **22.** ____________

23. Which year had the largest increase over the previous year? **23.** ____________

24. Which year had the highest net sales? **24.** ____________

25. Between which two years was the increase smallest? **25.** ____________

Name: Date:
Instructor: Section:

The line graph shows the annual sales for two different stores from 1996 to 2000. Use the line graph to answer exercises 26–30.

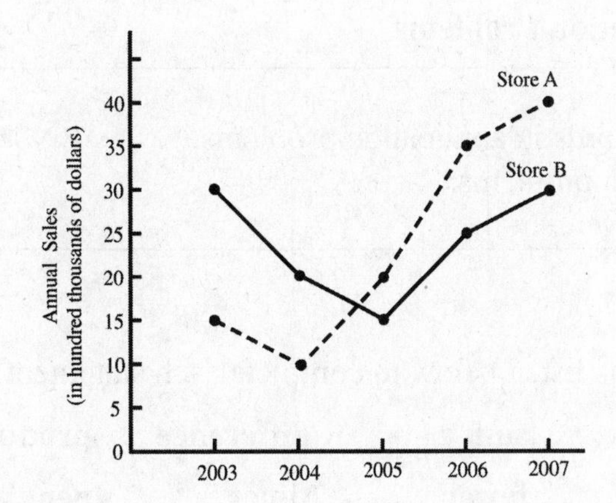

26. In which years did the sales of store A exceed the sales of store B? **26.** ______________

27. Which year showed the least difference between the sales of store A and the sales of store B? **27.** ______________

28. Which year showed the greatest difference between the sales of store A and the sales of store B? **28.** ______________

29. What was the difference in annual sales between store A and store B in 1996? **29.** ______________

30. Between which two years was the increase for both stores greatest? **30.** ______________

Name: Date:
Instructor: Section:

Chapter 1 WHOLE NUMBERS

1.10 Solving Application Problems

Learning Objectives
1. Find indicator words in application problems.
2. Solve application problems.
3. Estimate an answer.

Key Terms

Use the vocabulary terms listed below to complete each statement in exercises 1–5.

indicator words	**sum**	**difference**	**product**	**quotient**
increased by	**fewer**	**times**	**per**	

1. Words in a problem that indicate the necessary operations are ______________________.

2. ______________________ and ______________________ are indicator words for addition.

3. ______________________ and ______________________ are indicator words for multiplication.

4. ______________________ and ______________________ are indicator words for division.

5. ______________________ and ______________________ are indicator words for subtraction.

Objective 1 Find indicator words in application problems.

Write the operation determined by each of the following.

1. decreased by **1.** ________________

2. more than **2.** ________________

3. twice **3.** ________________

4. goes into **4.** ________________

5. loss of **5.** ________________

Name: Date:
Instructor: Section:

Objective 2 Solve application problems.

Solve each application problem.

6. A truck weights 8950 pounds when empty. After being loaded with firewood, it weighs 17,180 pounds. What is the weight of the firewood?

6. ______________

7. How many 3-inch strips of leather can be cut from a piece of leather 1 foot wide? (Hint: 1 foot = 12 inches.)

7. ______________

8. If there are 43,560 square feet in an acre, how many square feet are there in 5 acres?

8. ______________

9. Travel Rent-A-Car owns 365 compact cars, 438 full-sized cars, 125 luxury cars, and 83 vans and trucks. How many vehicles does it have in all?

9. ______________

10. Amanda Raymond owes $5520 on a loan. Find her monthly payment if the loan is paid off in 48 months.

10. ______________

11. Total receipts at a concert were $191,800. Each ticket cost $28. How many people attended the concert?

11. ______________

12. Two sisters share a legal bill of $1903. One sister pays $954 toward the bill. How much must the other sister pay?

12. ______________

Name: Date:
Instructor: Section:

13. A biology class found 14 deer in one area, 158 in another, and 417 in a third. How many deer did the class find?

13. ______________

14. Baseball uniforms cost $79 each. Find the cost of 23 uniforms.

14. ______________

15. The number of gallons of water polluted each day in an industrial area is 219,530. How many gallons are polluted each year? (Use a 365-day year.)

15. ______________

16. Shari bought 3 books costing $12, $17, and $18 each. She paid with a $50 bill. How much change will she receive?

16. ______________

17. A new car costs $11,350 before a trade-in. The car can be paid off in 36 monthly payments of $209 each after the trade-in. Find the amount of the trade-in.

17. ______________

18. Blue Bird leader, Barbara Walton, estimates that each of her Blue Birds will eat 2 cookies while she and her assistant, Lana Meehan, will eat 3 cookies each. If she expects 15 Blue Birds and her assistant at the meeting, how many cookies will she need?

18. ______________

Name: Date:
Instructor: Section:

19. Edward Biondi has $3117 in his checking account. If he pays $340 for tires, $725 for equipment repairs, and $198 for fuel and oil, find the balance remaining in his account.

19. ______________

20. Rodney Guess owns 55 acres of land which he leases to an alfalfa farmer for $150 per acre per year. If property taxes are $28 per acre per year, find the total amount he has left after taxes are paid.

20. ______________

Objective 3 Estimate an answer.

First use front end rounding to estimate the answer. Then find the exact answer.

21. A bus traveled 605 miles at 55 miles per hour. How long did the trip take?

21.
Estimate ______________
Exact ______________

22. Liz Skinner has $3712 in her checking account. After writing a check of $887 for tuition and parking fees, how much remains in her account?

22.
Estimate ______________
Exact ______________

23. Lori Knight knows that her car gets 36 miles per gallon in town. How many miles can she travel on 26 gallons?

23.
Estimate ______________
Exact ______________

24. Ski Mart offers a set of skis at a sale price of $219. If the sale price gives a savings of $56 off the original price, what is the original price of the skis?

24.
Estimate ______________
Exact ______________

Name: Date:
Instructor: Section:

25. If there are 43,560 square feet in an acre, how many square feet are there in 5 acres?

25.
Estimate____________

Exact ____________

26. If 560 stamps are divided evenly among 16 collectors, how many stamps will each receive?

26.
Estimate____________

Exact ____________

27. A room measures 18 feet by 12 feet. If carpeting costs $23 per square yard, find the total cost for carpeting the room.
(Hint: one square yard = 3 feet × 3 feet.)

27.
Estimate____________

Exact ____________

28. The Top Hat Grille finds that it needs five pounds of hamburger to make 35 servings of chili. How many pounds of hamburger are needed to make 182 servings of chili?

28.
Estimate____________

Exact ____________

29. Jerri Taft's vending machine company had 325 machines on hand at the beginning of the month. At different times during the month, machines were distributed to new locations: 37 machines were taken at one time, then 24 machines, and then 81 machines. During the same month additional machines were returned: 16 machines were returned at one time, then 39 machines, and then 110 machines. How many machines were on hand at the end of the month?

29.
Estimate____________

Exact ____________

30. Diana Ditka spent $286 on tuition, $137 on books, and $32 on supplies. If this money is withdrawn from her checking account, which had a balance of $723, what is her new balance.

30.
Estimate____________

Exact ____________

Name: Date:
Instructor: Section:

Chapter 2 MULTIPLYING AND DIVIDING FRACTIONS

2.1 Basics of Fractions

Learning Objectives
1. Use a fraction to show which part of a whole is shaded.
2. Identify the numerator and denominator.
3. Identify proper and improper fractions.

Key Terms

Use the vocabulary terms listed below to complete each statement in exercises 1–4.

numerator **denominator** **proper fraction** **improper fraction**

1. A fraction whose numerator is larger than its denominator is called an ____________________.

2. In the fraction $\frac{2}{9}$, the 2 is the ____________________.

3. A fraction whose denominator is larger than its numerator is called a ____________________.

4. The ____________________ of a fraction shows the number of equal parts in a whole.

Objective 1 Use a fraction to show which part of a whole is shaded.

Write the fractions that represent the shaded and unshaded portions of each figure.

1. 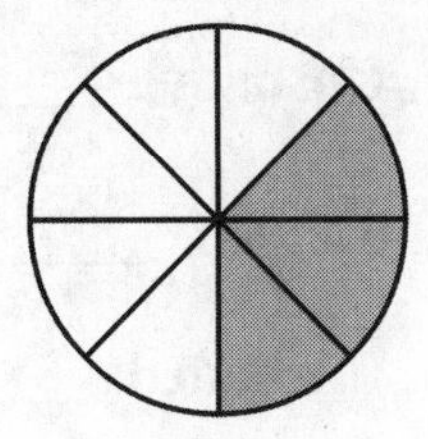

1. Shaded __________

Unshaded __________

2. 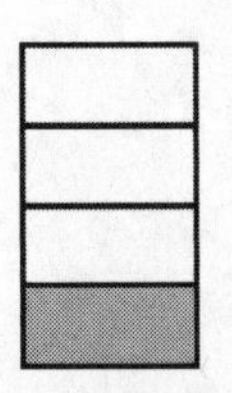

2. Shaded __________

Unshaded __________

3. 

3. Shaded __________

Unshaded __________

Name: Date:
Instructor: Section:

4. 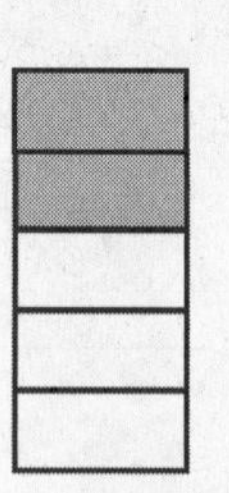

4. Shaded ________

Unshaded ________

5. 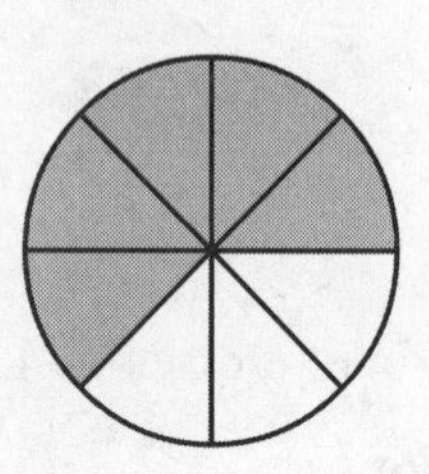

5. Shaded ________

Unshaded ________

6. 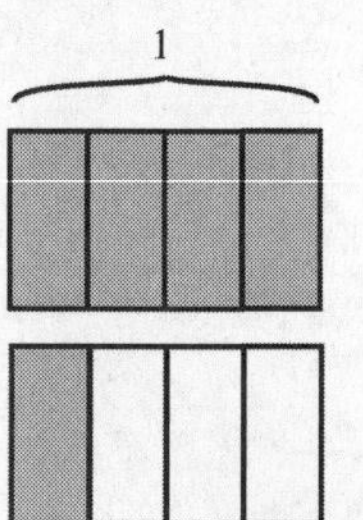

6. Shaded ________

Unshaded ________

7. 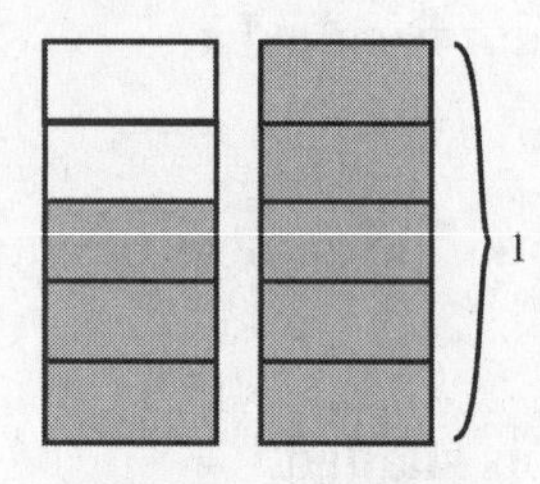

7. Shaded ________

Unshaded ________

8. 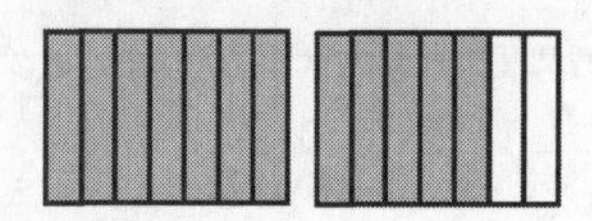

8. Shaded ________

Unshaded ________

9. 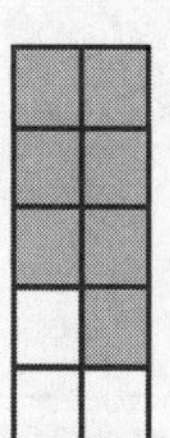

9. Shaded ________

Unshaded ________

10. 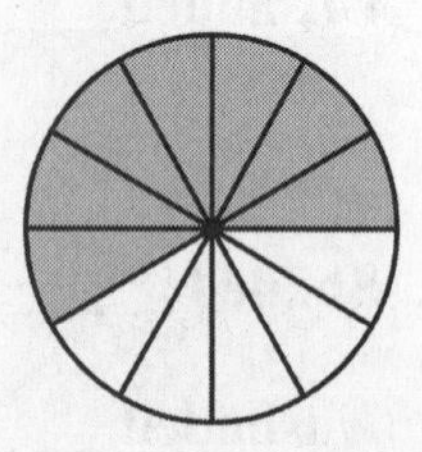

10. Shaded ________

Unshaded ________

Name: Date:
Instructor: Section:

Objective 2 Identify the numerator and denominator.

Identify the numerator and denominator.

11. $\frac{4}{3}$

11.
Numerator __________

Denominator ________

12. $\frac{2}{5}$

12.
Numerator __________

Denominator ________

13. $\frac{8}{11}$

13.
Numerator __________

Denominator ________

14. $\frac{11}{8}$

14.
Numerator __________

Denominator ________

15. $\frac{19}{50}$

15.
Numerator __________

Denominator ________

16. $\frac{112}{5}$

16.
Numerator __________

Denominator ________

17. $\frac{19}{8}$

17.
Numerator __________

Denominator ________

18. $\frac{98}{13}$

18.
Numerator __________

Denominator ________

19. $\frac{157}{12}$

19.
Numerator __________

Denominator ________

Name: Date:
Instructor: Section:

20. $\frac{14}{195}$

20.
Numerator ____________

Denominator ____________

Objective 3 Identify proper and improper fractions.

Write whether each fraction is **proper** *or* **improper**.

21. $\frac{9}{7}$ **21.** ________________

22. $\frac{5}{12}$ **22.** ________________

23. $\frac{7}{15}$ **23.** ________________

24. $\frac{17}{11}$ **24.** ________________

25. $\frac{4}{19}$ **25.** ________________

26. $\frac{11}{7}$ **26.** ________________

27. $\frac{18}{18}$ **27.** ________________

28. $\frac{7}{12}$ **28.** ________________

29. $\frac{3}{4}$ **29.** ________________

30. $\frac{10}{10}$ **30.** ________________

Name: Date:
Instructor: Section:

Chapter 2 MULTIPLYING AND DIVIDING FRACTIONS

2.2 Mixed Numbers

Learning Objectives

1. Identify mixed numbers.
2. Write mixed numbers as improper fractions.
3. Write improper fractions as mixed numbers.

Key Terms

Use the vocabulary terms listed below to complete each statement in exercises 1–4.

mixed number **improper fraction** **proper fraction**

whole numbers

1. The fraction $\frac{5}{8}$ is an example of a ______________________.

2. A ______________________ includes a fraction and a whole number written together.

3. A mixed number can be rewritten as an ______________________.

4. 0, 1, 2, 3, … are ______________________.

Objective 1 Identify mixed numbers.

List the mixed numbers in each group.

1. $2\frac{1}{2}, \frac{3}{5}, 1\frac{1}{6}, \frac{3}{4}$ **1.** ______________

2. $\frac{3}{8}, 5\frac{2}{3}, \frac{7}{4}, 3\frac{1}{2}$ **2.** ______________

3. $\frac{8}{7}, \frac{10}{10}, \frac{2}{3}, \frac{0}{5}$ **3.** ______________

4. $\frac{9}{9}, 3\frac{1}{2}, 10\frac{1}{3}, \frac{8}{2}, \frac{7}{9}$ **4.** ______________

5. $\frac{6}{3}, 4\frac{3}{4}, \frac{10}{10}, \frac{1}{3}, \frac{0}{8}$ **5.** ______________

Name: Date:
Instructor: Section:

Objective 2 Write mixed numbers as improper fractions.

Write each mixed number as an improper fraction.

6. $2\frac{7}{8}$ 6. ________

7. $1\frac{5}{6}$ 7. ________

8. $2\frac{4}{5}$ 8. ________

9. $5\frac{4}{7}$ 9. ________

10. $1\frac{3}{4}$ 10. ________

11. $6\frac{1}{4}$ 11. ________

12. $4\frac{2}{3}$ 12. ________

13. $7\frac{1}{2}$ 13. ________

14. $2\frac{7}{11}$ 14. ________

15. $5\frac{3}{7}$ 15. ________

16. $6\frac{2}{3}$ 16. ________

17. $8\frac{7}{9}$ 17. ________

18. $13\frac{3}{9}$ 18. ________

Name: Date:
Instructor: Section:

Objective 3 Write improper fractions as mixed numbers.

Write each improper fraction as a mixed number.

19. $\frac{14}{9}$ **19.** ____________

20. $\frac{20}{7}$ **20.** ____________

21. $\frac{29}{9}$ **21.** ____________

22. $\frac{26}{7}$ **22.** ____________

23. $\frac{21}{5}$ **23.** ____________

24. $\frac{41}{9}$ **24.** ____________

25. $\frac{25}{9}$ **25.** ____________

26. $\frac{29}{4}$ **26.** ____________

27. $\frac{92}{3}$ **27.** ____________

28. $\frac{211}{11}$ **28.** ____________

29. $\frac{749}{17}$ **29.** ____________

30. $\frac{2573}{11}$ **30.** ____________

Name: Date:
Instructor: Section:

Chapter 2 MULTIPLYING AND DIVIDING FRACTIONS

2.3 Factors

Learning Objectives
1 Find factors of a number.
2 Identify prime numbers
3 Find prime factorizations.

Key Terms

Use the vocabulary terms listed below to complete each statement in exercises 1–5.

factors **composite number** **prime number**

factorizations **prime factorization**

1. The numbers that can be multiplied to give a specific number (product) are ____________________ of that number.

2. A ________________________ has at least one factor other than itself and 1.

3. In a ________________________________ every factor is a prime number.

4. The factors of a ______________________ are itself and 1.

5. Numbers that are multiplied to give a product are ______________________.

Objective 1 Find factors of a number.

Find all the factors of each number.

1. 7 — 1. ______________

2. 12 — 2. ______________

3. 49 — 3. ______________

4. 15 — 4. ______________

5. 10 — 5. ______________

6. 36 — 6. ______________

7. 25 **7.** ______________

8. 24 **8.** ______________

9. 30 **9.** ______________

10. 72 **10.** ______________

Objective 2 Identify prime numbers.

Write whether each number is **prime**, **composite**, *or* **neither**.

11. 1 **11.** ______________

12. 5 **12.** ______________

13. 15 **13.** ______________

14. 11 **14.** ______________

15. 24 **15.** ______________

16. 45 **16.** ______________

17. 2 **17.** ______________

18. 31 **18.** ______________

19. 29 **19.** ______________

20. 38 **20.** ______________

Objective 3 Find prime factorizations.

Find the prime factorization of each number. Write the answer with exponents when repeated factors appear.

21. 12 **21.** ______________

Name: Date:
Instructor: Section:

22. 27 **22.** ________________

23. 28 **23.** ________________

24. 42 **24.** ________________

25. 24 **25.** ________________

26. 72 **26.** ________________

27. 108 **27.** ________________

28. 160 **28.** ________________

29. 450 **29.** ________________

30. 171 **30.** ________________

Name: Date:
Instructor: Section:

Chapter 2 MULTIPLYING AND DIVIDING FRACTIONS

2.4 Writing a Fraction in Lowest Terms

Learning Objectives

1. Tell whether a fraction is written in lowest terms.
2. Write a fraction in lowest terms using common factors.
3. Write a fraction in lowest terms using prime factors.
4. Determine whether two fractions are equivalent.

Key Terms

Use the vocabulary terms listed below to complete each statement in exercises 1–3.

equivalent fractions **common factor** **lowest terms**

1. A fraction is written in ______________________________ when its numerator and denominator have no common factor other than 1.

2. A ________________________ is a number that can be divided into two or more whole numbers.

3. Two fractions are ______________________________ when they represent the same portion of a whole.

Objective 1 Tell whether a fraction is written in lowest terms.

Write whether or not each fraction is in lowest terms. Write **yes** *or* **no**.

1. $\frac{4}{12}$ **1.** ________________

2. $\frac{3}{7}$ **2.** ________________

3. $\frac{12}{18}$ **3.** ________________

4. $\frac{13}{17}$ **4.** ________________

5. $\frac{7}{19}$ **5.** ________________

6. $\frac{3}{39}$ **6.** ________________

Name: Date:
Instructor: Section:

Objective 2 Write a fraction in lowest terms using common factors.

Write each fraction in lowest terms.

7. $\frac{4}{12}$ **7.** ________

8. $\frac{14}{49}$ **8.** ________

9. $\frac{8}{36}$ **9.** ________

10. $\frac{26}{39}$ **10.** ________

11. $\frac{28}{98}$ **11.** ________

12. $\frac{30}{42}$ **12.** ________

13. $\frac{12}{88}$ **13.** ________

14. $\frac{16}{56}$ **14.** ________

Name: Date:
Instructor: Section:

Objective 3 Write a fraction in lowest terms using prime factors.

Write the numerator and denominator of each fraction as a product of prime factors and divide by the common factors. Then write the fraction in lowest terms.

15. $\frac{63}{84}$

15. ________________

16. $\frac{28}{56}$

16. ________________

17. $\frac{180}{210}$

17. ________________

18. $\frac{72}{90}$

18. ________________

19. $\frac{36}{54}$

19. ________________

20. $\frac{71}{142}$

20. ________________

21. $\frac{75}{500}$

21. ________________

22. $\frac{96}{132}$

22. ________________

Name: Date:
Instructor: Section:

Objective 4 Determine whether two fractions are equivalent.

Determine whether each pair of fractions is **equivalent** *or* **not equivalent.**

23. $\frac{3}{7}$ and $\frac{6}{14}$

23. ______________

24. $\frac{2}{3}$ and $\frac{10}{15}$

24. ______________

25. $\frac{6}{21}$ and $\frac{3}{7}$

25. ______________

26. $\frac{8}{16}$ and $\frac{15}{20}$

26. ______________

27. $\frac{9}{12}$ and $\frac{6}{8}$

27. ______________

28. $\frac{12}{28}$ and $\frac{18}{42}$

28. ______________

29. $\frac{20}{24}$ and $\frac{15}{31}$

29. ______________

30. $\frac{6}{12}$ and $\frac{8}{16}$

30. ______________

Name: Date:
Instructor: Section:

Chapter 2 MULTIPLYING AND DIVIDING FRACTIONS

2.5 Multiplying Fractions

Learning Objectives	
1	Multiply fractions.
2	Use a multiplication shortcut.
3	Multiply a fraction and a whole number.
4	Find the area of a rectangle.

Key Terms

Use the vocabulary terms listed below to complete each statement in exercises 1–4.

multiplication shortcut **numerator** **denominator**

common factor

1. A ____________________ can be divided into two or more whole numbers.

2. The number below the fraction bar in a fraction is called the ________________.

3. When multiplying fractions, the process of dividing a numerator and denominator by a common factor can be used as a ____________________________________.

4. The number above the fraction bar in a fraction is called the ________________.

Objective 1 Multiply factions.

Multiply. Write answers in lowest terms.

1. $\frac{5}{9} \cdot \frac{7}{6}$ **1.** ________________

2. $\frac{4}{7} \cdot \frac{3}{5}$ **2.** ________________

3. $\frac{5}{6} \cdot \frac{11}{4}$ **3.** ________________

4. $\frac{9}{10} \cdot \frac{3}{2}$ **4.** ________________

5. $\frac{3}{4} \cdot \frac{5}{6} \cdot \frac{2}{3}$ **5.** ________________

Name: Date:
Instructor: Section:

6. $\frac{1}{9} \cdot \frac{2}{3} \cdot \frac{5}{6}$

6. ____________

7. $\frac{3}{8} \cdot \frac{1}{4} \cdot \frac{1}{9}$

7. ____________

Objective 2 Use a multiplication shortcut.

Use the multiplication shortcut to find each product. Write the answer in lowest terms.

8. $\frac{7}{6} \cdot \frac{3}{14}$

8. ____________

9. $\frac{4}{9} \cdot \frac{15}{16}$

9. ____________

10. $\frac{3}{5} \cdot \frac{25}{27}$

10. ____________

11. $\frac{11}{4} \cdot \frac{8}{33}$

11. ____________

12. $\frac{5}{6} \cdot \frac{4}{35}$

12. ____________

13. $\frac{3}{4} \cdot \frac{5}{9} \cdot \frac{2}{5}$

13. ____________

14. $\frac{3}{8} \cdot \frac{4}{9} \cdot \frac{15}{6}$

14. ____________

15. $\frac{25}{35} \cdot \frac{14}{30} \cdot \frac{3}{7}$

15. ____________

Name: Date:
Instructor: Section:

Objective 3 Multiply a fraction and a whole number.

Multiply. Write the answer in lowest terms. Change the answer to a whole or mixed number where possible.

16. $6 \cdot \frac{7}{300}$ **16.** ______________

17. $\frac{4}{250} \cdot 50$ **17.** ______________

18. $27 \cdot \frac{7}{54}$ **18.** ______________

19. $49 \cdot \frac{6}{7}$ **19.** ______________

20. $21 \cdot \frac{3}{7} \cdot \frac{7}{9}$ **20.** ______________

21. $\frac{9}{26} \cdot \frac{39}{18} \cdot 12$ **21.** ______________

22. $200 \cdot \frac{7}{50} \cdot \frac{5}{28}$ **22.** ______________

23. $\frac{21}{520} \cdot 13 \cdot \frac{20}{7}$ **23.** ______________

Objective 4 Find the area of a rectangle.

Find the area of each rectangle.

24. Length: $\frac{2}{3}$ yard, width: $\frac{1}{2}$ yard **24.** ______________

Name: Date:
Instructor: Section:

25. $\frac{3}{4}$ meter, width: $\frac{1}{2}$ meter **25.** ________________

26. Length: $\frac{5}{3}$ yards, width: $\frac{3}{2}$ yards **26.** ________________

27. Length: $\frac{9}{16}$ meter, width: $\frac{5}{6}$ meter **27.** ________________

28. Length: $\frac{7}{16}$ inch, width: $\frac{3}{16}$ inch **28.** ________________

Solve each application problem.

29. A desk is $\frac{2}{3}$ yard by $\frac{5}{6}$ yard. Find its area. **29.** ________________

30. A wading pool is $\frac{5}{4}$ yards by $\frac{5}{9}$ yard. Find its area **30.** ________________

Name: Date:
Instructor: Section:

Chapter 2 MULTIPLYING AND DIVIDING FRACTIONS

2.6 Applications of Multiplication

Learning Objectives
1 Solve fraction application problems using multiplication.

Key Terms

Use the vocabulary terms listed below to complete each statement in exercises 1–3.

reciprocal **product** **indicator words**

1. The words "times" and "double" are ____________________ for multiplication.
2. In the problem $51 \times 3 = 153$, 153 is called the ____________________.
3. Two numbers are ________________________ of each other if their product is 1.

Objective 1 Solve fraction application problems using multiplication.

Solve each application problem.

1. A bookstore sold 2800 books, $\frac{3}{5}$ of which were paperbacks. How many paperbacks were sold? 1. ________________

2. A store sells 3750 items, of which $\frac{2}{15}$ are classified as junk food. How many of the items are junk food? 2. ________________

3. Sara needs $2500 to go to school for one year. She earns $\frac{3}{5}$ of this amount in the summer. How much does she earn in the summer? 3. ________________

4. Lani paid $120 for textbooks this term. Of this amount, the bookstore kept $\frac{1}{4}$. How much did the bookstore keep? 4. ________________

Name: Date:
Instructor: Section:

5. Of the 570 employees of Grand Tire Service, $\frac{7}{30}$ have given to the United Fund. How many have given to the United Fund? **5.** ______________

6. A school gives scholarships to $\frac{3}{25}$ of its 1900 freshmen. How many students receive scholarships? **6.** ______________

7. Kim earns \$1500 a month. If she uses $\frac{1}{3}$ of her income on housing, how much does she pay for housing? **7.** ______________

8. Akiko's home is $\frac{3}{5}$ of the way from Carolyn's home to Laplace College, a distance of 45 miles. How far is it from Carolyn's home to Akiko's? **8.** ______________

9. Deepak puts $\frac{1}{12}$ of his weekly earnings in a retirement fund. If he makes \$1248 a week, how much does he put in his retirement fund each week? **9.** ______________

10. The sophomore class at Lincoln High School has 312 students. If $\frac{7}{13}$ of the students are boys, how many boys are in the sophomore class? **10.** ______________

Name: Date:
Instructor: Section:

11. The Donut Shack sells donuts, bagels, and muffins. During a typical week, they sell 1120 items, of which $\frac{2}{7}$ are muffins. How many muffins does the Donut Shack sell in a typical week?

11. ________________

12. During the month of February $\frac{5}{7}$ of the days had temperatures that were below normal. How many days had below normal temperatures? (Assume that this is not a leap year)

12. ________________

13. Marcie is reading a 360 page book. How many pages has she read if she had completed $\frac{4}{9}$ of the book?

13. ________________

14. A local hospital is recruiting new blood donors. During the month of June, $\frac{3}{16}$ of the people who donated were first-time donors. If 112 people donated blood in June, how many were first time donors?

14. ________________

15. Major league baseball teams play 162 games during the regular season. If a team wins $\frac{15}{27}$ of its games, how many games does it win?

15. ________________

16. A lawnmower uses a gasoline/oil mixture in which $\frac{1}{30}$ of the mixture must be oil. If the tank holds 150 ounces of the mixture, how many ounces are oil?

16. ________________

Name: Date:
Instructor: Section:

17. During a local election, a candidate received $\frac{2}{3}$ of the votes. If 2532 people voted in the election, how many votes did the candidate receive?

17. ______________

18. George takes a train to work everyday. The distance from George's house to the train station is $\frac{1}{8}$ of the total distance of 48 miles to work. What is the distance from his home to the train station?

18. ______________

19. Mary must calculate the area of her home office for tax purposes. What is the area of Mary's home office if it takes up $\frac{2}{15}$ of her entire 2340 square foot home?

19. ______________

20. Darrell is participating in a 150-mile bike ride for charity. There are 10 equally-spaced stops including the finish line. How many miles has Darrell completed after reaching the 7th stop?

20. ______________

21. Elena noticed that her gas gauge moved from the $\frac{7}{8}$ mark to the $\frac{4}{8}$ mark during a recent trip. If her tank holds 16 gallons, how many gallons did she use during this trip?

21. ______________

22. A local college has 8700 undergraduate students and $\frac{7}{30}$ of these students commute to school. How many students commute to school?

22. ______________

Name: Date:
Instructor: Section:

The Hu family earned $54,000 last year. Use this fact to solve Problems 23–26.

23. They paid $\frac{1}{3}$ of their income for taxes. How much did they pay in taxes?

23. ________________

24. They spend $\frac{2}{5}$ of their income for rent. How much did they spend on rent?

24. ________________

25. They saved $\frac{1}{16}$ of their income. How much did they save?

25. ________________

26. They spend $\frac{1}{6}$ of their income on food. How much did they spend on food?

26. ________________

The Highview Condo Association collects $96,000 each year from its members. Use this fact to solve Problems 27–30.

27. The association spends $\frac{1}{6}$ of this money on landscaping. How much do they spend on landscaping?

27. ________________

28. The association spends $\frac{5}{12}$ of this money on routine maintenance. How much do they spend on routine maintenance?

28. ________________

29. The association spends $\frac{3}{8}$ on insurance. How much to they spend on insurance?

29. ________________

Name: Date:
Instructor: Section:

30. The association puts $\frac{3}{32}$ of this money in an emergency fund. How much money do they put in this fund?

30. ______________

Name: Date:
Instructor: Section:

Chapter 2 MULTIPLYING AND DIVIDING FRACTIONS

2.7 Dividing Fractions

Learning Objectives	
1	Find the reciprocal of a fraction.
2	Divide fractions.
3	Solve application problems in which fractions are divided.

Key Terms

Use the vocabulary terms listed below to complete each statement in exercises 1–3.

reciprocals **indicator words** **quotient**

1. Two numbers are ______________________ of each other if their product is 1.

2. The words "per" and "divided equally" are ______________________ for division.

3. In the problem $192 \div 12 = 16$, 16 is called the ____________________.

Objective 1 Find the reciprocal of a fraction.

Find the reciprocal of each fraction.

1. $\frac{3}{4}$ **1.** ______________

2. $\frac{9}{2}$ **2.** ______________

3. $\frac{1}{3}$ **3.** ______________

4. $\frac{6}{7}$ **4.** ______________

5. 10 **5.** ______________

6. $\frac{15}{4}$ **6.** ______________

Name: Date:
Instructor: Section:

Objective 2 Divide fractions

Divide. Write the answer in lowest terms. Change the answers to a whole or mixed number where possible.

7. $\frac{4}{5} \div \frac{3}{8}$

7. ____________

8. $\frac{28}{5} \div \frac{42}{25}$

8. ____________

9. $\frac{\frac{7}{10}}{\frac{14}{5}}$

9. ____________

10. $\frac{\frac{4}{9}}{\frac{16}{27}}$

10. ____________

11. $9 \div \frac{3}{2}$

11. ____________

12. $\frac{5}{8} \div 15$

12. ____________

13. $\frac{\frac{11}{3}}{5}$

13. ____________

14. $\frac{\frac{6}{11}}{18}$

14. ____________

15. $\frac{\frac{8}{15}}{\frac{10}{12}}$

15. ____________

16. $4 \div \frac{12}{7}$

16. ____________

Name: Date:
Instructor: Section:

17. $\frac{3}{4} \div \frac{27}{8}$ **17.** ______________

18. $\frac{75}{8} \div 16$ **18.** ______________

19. $16 \div \frac{75}{8}$ **19.** ______________

Objective 3 Solve application problems in which fractions are divided.

Solve each application problem.

20. Abel has a piece of property with an area of $\frac{7}{8}$ acre. He wishes to divide it into four equal parts for his children. How many acres of land will each child get? **20.** ______________

21. Amanda wants to make doll dresses to sell at a craft's fair. Each dress needs $\frac{1}{3}$ yard of material. She has 18 yards of material. Find the number of dresses that she can make. **21.** ______________

22. It takes $\frac{4}{5}$ pound of salt to fill a large salt shaker. How many salt shakers can be filled with 32 pounds of salt? **22.** ______________

23. Lynn has 2 gallons of lemonade. If each of her Brownies gets $\frac{1}{12}$ gallon of lemonade, how many Brownies does she have? **23.** ______________

Name: Date:
Instructor: Section:

24. How many $\frac{1}{9}$-ounce medicine vials can be filled with 7 ounces of medicine? **24.** ______________

25. Each guest at a party will eat $\frac{5}{16}$ pound of chips. How many guests can be served with 10 pounds of chips? **25.** ______________

26. Samantha uses $\frac{2}{3}$ yard of ribbon to make a bow for each package she wraps at May's Department Store. How many bows can she make if she has 60 yards of ribbon? **26.** ______________

27. Bill wishes to make hamburger patties that weight $\frac{5}{12}$ pound. How many hamburger patties can he make with 10 pounds of hamburger? **27** ______________

28. Glen has a small pickup truck that will carry $\frac{3}{4}$ cord of firewood. Find the number of trips needed to deliver 30 cords of wood. **28.** ______________

29. How many $\frac{5}{4}$-cup glass tumblers can be filled from a 20-cup bowl of punch? **29.** ______________

30. Janine wants to make wooden coasters for glasses. How many coasters can she make from a 12-inch round post if each coaster is to be $\frac{3}{8}$-inch tall? **30.** ______________

Name: Date:
Instructor: Section:

Chapter 2 MULTIPLYING AND DIVIDING FRACTIONS

2.8 Multiplying and Dividing Mixed Numbers

Learning Objectives	
1	Estimate the answer and multiply mixed numbers.
2	Estimate the answer and divide mixed numbers.
3	Solve application problems with mixed numbers.

Key Terms

Use the vocabulary terms listed below to complete each statement in exercises 1–3.

mixed number **simplify** **round**

1. To ____________________ a fraction means to write the fraction in lowest terms.

2. If the numerator of a fraction is half of the denominator or more, ____________ up to the next whole number to estimate the product of a mixed number and a whole number.

3. $2\frac{7}{11}$ is an example of a ____________________.

Objective 1 Estimate the answer and multiply mixed numbers.

First estimate the answer. Then multiply to find the exact answer. Simplify all answers.

1. $5\frac{1}{3} \cdot 2\frac{1}{2}$

   1. **Estimate** ____________

   Exact ____________

2. $3\frac{1}{2} \cdot 4\frac{2}{7}$

   2. **Estimate** ____________

   Exact ____________

3. $3\frac{1}{2} \cdot 1\frac{3}{7}$

   3. **Estimate** ____________

   Exact ____________

4. $4\frac{4}{9} \cdot 2\frac{2}{5}$

   4. **Estimate** ____________

   Exact ____________

Name: Date:
Instructor: Section:

5. $5\frac{2}{3}\cdot 7\frac{1}{8}$

5.
Estimate________

Exact ________

6. $18\cdot 2\frac{5}{9}$

6.
Estimate________

Exact ________

7. $3\frac{2}{5}\cdot 15$

7.
Estimate________

Exact ________

8. $\frac{5}{6}\cdot 2\frac{1}{2}\cdot 2\frac{2}{5}$

8.
Estimate________

Exact ________

9. $1\frac{1}{4}\cdot 1\frac{1}{3}\cdot 1\frac{1}{2}$

9.
Estimate________

Exact ________

10. $9\cdot 3\frac{1}{4}\cdot 1\frac{3}{13}\cdot 2\frac{2}{3}$

10.
Estimate________

Exact ________

Objective 2 Estimate the answer and divide mixed numbers.

First estimate the answer. Then divide to find the exact answer. Simplify all answers.

11. $5\frac{5}{6}\div 5\frac{1}{4}$

11.
Estimate________

Exact ________

12. $4\frac{5}{8}\div 1\frac{1}{4}$

12.
Estimate________

Exact ________

13. $4\frac{3}{8}\div 3\frac{1}{2}$

13.
Estimate________

Exact ________

Name: Date:
Instructor: Section:

14. $6\frac{1}{4} \div 2\frac{1}{2}$

14.
Estimate ____________

Exact ____________

15. $14 \div 8\frac{2}{5}$

15.
Estimate ____________

Exact ____________

16. $7\frac{1}{3} \div 6$

16.
Estimate ____________

Exact ____________

17. $7\frac{1}{2} \div \frac{2}{3}$

17.
Estimate ____________

Exact ____________

18. $2\frac{5}{8} \div 1\frac{3}{4}$

18.
Estimate ____________

Exact ____________

19. $8\frac{3}{4} \div 5$

19.
Estimate ____________

Exact ____________

20. $16 \div 2\frac{7}{8}$

20.
Estimate ____________

Exact ____________

Objective 3 Solve application problems with mixed numbers.

First estimate the answer. Then solve each application problem. Simplify all answers.

21. Maria wants to make 20 dresses to sell at a bazaar. Each dress needs $3\frac{1}{4}$ yards of material. How many yards does she need?

21.
Estimate ____________

Exact ____________

Name: Date:
Instructor: Section:

22. Juan worked $38\frac{1}{4}$ hours at $9 per hour. How much did he make?

22.
Estimate______________

Exact ______________

23. Each home in an area needs $41\frac{1}{3}$ yards of rain gutter. How much rain gutter would be needed for 6 homes?

23.
Estimate______________

Exact ______________

24. A farmer applies fertilizer to this fields at a rate of $5\frac{5}{6}$ gallons per acre. How many acres can he fertilize with $65\frac{5}{6}$ gallons?

24.
Estimate______________

Exact ______________

25. Insect spray is mixed using $1\frac{3}{4}$ ounces of a chemical per gallon of water. How many ounces of the chemical are needed to mix with $28\frac{4}{5}$ gallons of water?

25.
Estimate______________

Exact ______________

26. How many $\frac{3}{4}$-pound peanut cans can be filled with 15 pounds of peanuts?

26.
Estimate______________

Exact ______________

27. How many dresses can be made from 70 yards of material if each dress requires $4\frac{3}{8}$ yards?

27.
Estimate______________

Exact ______________

28. Arnette worked $24\frac{1}{2}$ hours and earned $9 per hour. How much did she earn?

28.
Estimate______________

Exact ______________

Name: Date:
Instructor: Section:

29. Juan has $3\frac{1}{2}$ sticks of margarine. If each stick weighs $\frac{1}{4}$ pound, how much does Juan's margarine weigh?

29.
Estimate____________

Exact ____________

30. A dental office plays taped music constantly. Each tape takes $1\frac{1}{4}$ hours. How many tapes are played during $7\frac{1}{2}$ hours?

30.
Estimate____________

Exact ____________

Name: Date:
Instructor: Section:

Chapter 3 ADDING AND SUBTRACTING FRACTIONS

3.1 Adding and Subtracting Like Fractions

Learning Objectives

1. Define like and unlike fractions.
2. Add like fractions.
3. Subtract like fractions.

Key Terms

Use the vocabulary terms listed below to complete each statement in exercises 1–2.

like fractions **unlike fractions**

1. Fractions with different denominators are called ________________________.
2. Fractions with the same denominator are called ________________________.

Objective 1 Define like and unlike fractions.

Write **like** *or* **unlike** *for each set of fractions.*

1. $\frac{3}{5}, \frac{4}{10}$ 1. ______________
2. $\frac{9}{7}, \frac{2}{7}$ 2. ______________
3. $\frac{2}{5}, \frac{3}{5}$ 3. ______________
4. $\frac{2}{3}, \frac{3}{2}$ 4. ______________
5. $\frac{2}{15}, \frac{3}{15}, \frac{1}{5}$ 5. ______________
6. $\frac{18}{7}, \frac{21}{7}, \frac{7}{7}$ 6. ______________

Name: Date:
Instructor: Section:

Objective 2 Add like fractions.

Add and simplify the answer.

7. $\frac{5}{8}+\frac{1}{8}$ **7.** ________

8. $\frac{11}{15}+\frac{1}{15}$ **8.** ________

9. $\frac{7}{8}+\frac{5}{8}$ **9.** ________

10. $\frac{4}{3}+\frac{7}{3}$ **10.** ________

11. $\frac{11}{16}+\frac{7}{16}$ **11.** ________

12. $\frac{1}{6}+\frac{5}{6}$ **12.** ________

13. $\frac{1}{5}+\frac{2}{5}+\frac{4}{5}$ **13.** ________

14. $\frac{6}{10}+\frac{4}{10}+\frac{3}{10}$ **14.** ________

15. $\frac{67}{81}+\frac{29}{81}+\frac{12}{81}$ **15.** ________

Solve each application problem. Write answers in lowest terms.

16. Malika walked $\frac{3}{8}$ of a mile downhill and then $\frac{1}{8}$ of a mile along a creek. How far did she walk altogether? **16.** ________

Name: Date:
Instructor: Section:

17. Last month the Yee family paid $\frac{2}{11}$ of a debt. This month they paid an additional $\frac{5}{11}$ of the same debt. What fraction of the debt has been paid?

17. ______________

18. Brent painted $\frac{1}{6}$ of a house last week and another $\frac{3}{6}$ this week. How much of the house is painted?

18. ______________

Objective 3 Subtract like fractions.

Subtract and simplify the answer.

19. $\frac{3}{10}-\frac{1}{10}$

19. ______________

20. $\frac{11}{16}-\frac{3}{16}$

20. ______________

21. $\frac{16}{21}-\frac{2}{21}$

21. ______________

22. $\frac{16}{15}-\frac{6}{15}$

22. ______________

23. $\frac{25}{28}-\frac{15}{28}$

23. ______________

24. $\frac{31}{36}-\frac{11}{36}$

24. ______________

25. $\frac{91}{100}-\frac{41}{100}$

25. ______________

Name: Date:
Instructor: Section:

26. $\frac{736}{400} - \frac{496}{400}$ **26.** ______________

27. $\frac{365}{224} - \frac{269}{224}$ **27.** ______________

Solve each application problem. Write answers in lowest terms.

28. Bill must walk $\frac{9}{12}$ of a mile. He has already walked $\frac{1}{12}$ of a mile. How much farther must he walk? **28.** ______________

29. Jeff planted $\frac{11}{18}$ of his garden in corn and potatoes. If $\frac{5}{18}$ of the garden is corn, how much of the garden is potatoes? **29.** ______________

30. The Thompsons owe $\frac{8}{15}$ of a debt. If they pay $\frac{2}{15}$ of it this month, what fraction of the debt will they still owe? **30.** ______________

Name: Date:
Instructor: Section:

Chapter 3 ADDING AND SUBTRACTING FRACTIONS

3.2 Least Common Multiples

Learning Objectives

1. Find the least common multiple.
2. Find the least common multiple using multiples of the largest number.
3. Find the least common multiple using prime factorization.
4. Find the least common multiple using an alternative method.
5. Write a fraction with an indicated denominator.

Key Terms

Use the vocabulary terms listed below to complete each statement in exercises 1–2.

least common multiple **LCM**

1. The ________________________________ of two whole numbers is the smallest whole number divisible by both of the numbers.

2. ______________ is the abbreviation for least common multiple.

Objective 1 Find the least common multiple.

Find the least common multiple for each of the following by listing multiples of each number.

1. 7, 14 **1.** ________________

2. 12, 18 **2.** ________________

3. 21, 28 **3.** ________________

4. 30, 75 **4.** ________________

5. 15, 21 **5.** ________________

Name: Date:
Instructor: Section:

Objective 2 Find the least common multiple using multiples of the largest number.

Find the least common multiple for each of the following by using multiples of the largest number.

6. 5, 12 **6.** ______________

7. 16, 20 **7.** ______________

8. 15, 25 **8.** ______________

9. 14, 35 **9.** ______________

10. 32, 40 **10.** ______________

Objective 3 Find the least common multiple using prime factorization.

Find the least common multiple for each of the following using prime factorization.

11. 14, 48 **11.** ______________

12. 28, 32 **12.** ______________

13. 10, 24, 32 **13.** ______________

14. 16, 20, 25 **14.** ______________

Name: Date:
Instructor: Section:

15. 7, 12, 21, 35 **15.** ________________

Objective 4 Find the least common multiple using an alternative method.

Find the least common multiple for each of the following using an alternative method.

16. 10, 15 **16.** ________________

17. 22, 55 **17.** ________________

18. 35, 85 **18.** ________________

19. 4, 18, 27 **19.** ________________

20. 12, 30, 40 **20.** ________________

21. 10, 12, 36 **21.** ________________

Objective 5 Write a fraction with an indicated denominator.

Rewrite each fraction with the indicated denominator.

22. $\frac{1}{9} = \frac{}{36}$ **22.** ________________

Name: Date:
Instructor: Section:

23. $\frac{2}{7} = \frac{}{63}$ **23.** ________

24. $\frac{1}{13} = \frac{}{78}$ **24.** ________

25. $\frac{5}{6} = \frac{}{72}$ **25.** ________

26. $\frac{3}{13} = \frac{}{52}$ **26.** ________

27. $\frac{21}{11} = \frac{}{55}$ **27.** ________

28. $\frac{15}{7} = \frac{}{84}$ **28.** ________

29. $\frac{9}{17} = \frac{}{102}$ **29.** ________

30. $\frac{7}{12} = \frac{}{60}$ **30.** ________

Name: Date:
Instructor: Section:

Chapter 3 ADDING AND SUBTRACTING FRACTIONS

3.3 Adding and Subtracting Unlike Fractions

Learning Objectives
1. Add unlike fractions.
2. Add unlike fractions vertically.
3. Subtract unlike fractions.
4. Subtract unlike fractions vertically.

Key Terms

Use the vocabulary terms listed below to complete each statement in exercises 1–2.

least common denominator **LCD**

1. In order to add or subtract fractions with different denominators, first find the
_______________________________________.

2. _______________ is the abbreviation for least common denominator.

Objective 1 Add unlike fractions.

Add the following fractions. Simplify all answers.

1. $\frac{1}{3}+\frac{1}{2}$ **1.** ________________

2. $\frac{1}{5}+\frac{5}{8}$ **2.** ________________

3. $\frac{3}{10}+\frac{7}{15}$ **3.** ________________

4. $\frac{1}{6}+\frac{3}{14}$ **4.** ________________

5. $\frac{4}{15}+\frac{9}{20}$ **5.** ________________

Name: Date:
Instructor: Section:

6. $\frac{1}{4}+\frac{2}{7}+\frac{3}{14}$ **6.** ______________

7. $\frac{1}{3}+\frac{1}{8}+\frac{5}{12}$ **7.** ______________

Solve each application problem. Write answers in lowest terms.

8. Michael Pippen paid $\frac{1}{9}$ of a debt in January, $\frac{1}{2}$ in February, $\frac{1}{4}$ in March, and $\frac{1}{12}$ in April. What fraction of the debt was paid in these four months? **8.** ______________

9. A buyer for a grain company bought $\frac{3}{8}$ ton of wheat, $\frac{1}{6}$ ton of rice, and $\frac{1}{4}$ ton of barley. How many tons of grain were bought? **9.** ______________

10. Find the perimeter (distance around) the figure. **10.** ______________

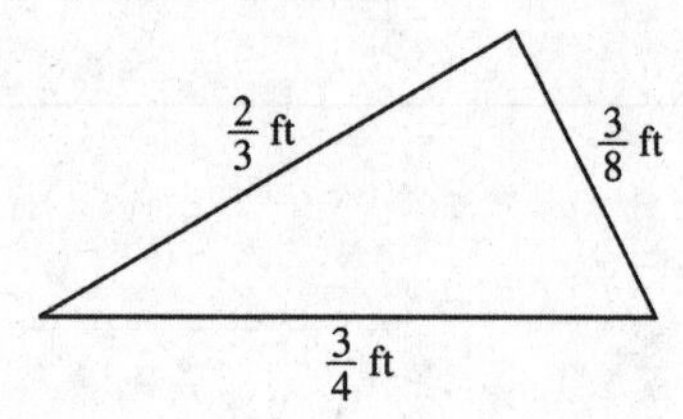

Objective 2 Add unlike fractions vertically.

Add the following fractions. Simplify all answers.

11. $\begin{array}{r} \frac{1}{15} \\ +\ \frac{2}{3} \\ \hline \end{array}$ **11.** ______________

Name: Date:
Instructor: Section:

12. $\frac{7}{12} + \frac{3}{8}$

12. ______________

13. $\frac{2}{15} + \frac{7}{10}$

13. ______________

14. $\frac{1}{6} + \frac{2}{9}$

14. ______________

15. $\frac{3}{7} + \frac{4}{21}$

15. ______________

16. $\frac{5}{22} + \frac{7}{33}$

16. ______________

17. $\frac{6}{13} + \frac{15}{52}$

17. ______________

Name: Date:
Instructor: Section:

18. $\begin{array}{r} \frac{3}{14} \\ + \frac{5}{21} \\ \hline \end{array}$

18. ______________

19. $\begin{array}{r} \frac{5}{18} \\ + \frac{7}{27} \\ \hline \end{array}$

19. ______________

20. $\begin{array}{r} \frac{1}{12} \\ + \frac{1}{8} \\ \hline \end{array}$

20. ______________

Objective 3 Subtract unlike fractions.
Objective 4 Subtract unlike fractions vertically.

Subtract the following fractions. Simplify all answers.

21. $\frac{7}{8} - \frac{1}{2}$

21. ______________

22. $\frac{5}{8} - \frac{1}{6}$

22. ______________

23. $\begin{array}{r} \frac{5}{9} \\ - \frac{5}{12} \\ \hline \end{array}$

23. ______________

Name: Date:
Instructor: Section:

24. $\begin{array}{r} \frac{9}{16} \\ -\frac{3}{10} \\ \hline \end{array}$

24. ______________

25. $\begin{array}{r} \frac{7}{8} \\ -\frac{7}{28} \\ \hline \end{array}$

25. ______________

26. $\frac{3}{5}-\frac{1}{4}$

26. ______________

27. $\frac{9}{10}-\frac{4}{25}$

27. ______________

Solve each application problem. Write answers in lowest terms.

28. A company has $\frac{5}{8}$ acre of land. They sold $\frac{1}{3}$ acre. How much land is left?

28. ______________

29. Greg had $\frac{7}{12}$ of his savings goal to complete at the beginning of the month. During the month he saved another $\frac{1}{8}$ of the goal. How much of the goal is left to save?

29. ______________

Name: Date:
Instructor: Section:

30. A $\frac{3}{4}$-inch nail was hammered through a board and $\frac{1}{8}$-inch of the nail stuck out. How thick is the board?

30. ________________

Name: Date:
Instructor: Section:

Chapter 3 ADDING AND SUBTRACTING FRACTIONS

3.4 Adding and Subtracting Mixed Numbers

Learning Objectives	
1	Estimate an answer, then add or subtract mixed numbers.
2	Estimate an answer, then subtract mixed numbers by regrouping.
3	Add or subtract mixed numbers using an alternate method.

Key Terms

Use the vocabulary terms listed below to complete each statement in exercises 1–2.

regrouping when adding fractions

regrouping when subtracting fractions

1. ________________________________ is the method used in the subtraction of mixed numbers when the fraction part of the minuend is less than the fraction part of the subtrahend.

2. ________________________________ is the method used in the addition of mixed numbers when the sum of the fraction is greater than 1.

Objective 1 Estimate an answer, then add or subtract mixed numbers.

First estimate the answer. Then add or subtract to find the exact answer. Write answers as mixed numbers in lowest terms.

1. $\begin{array}{r} 5\frac{1}{7} \\ +\ 4\frac{3}{7} \\ \hline \end{array}$

1. **Estimate** ____________ **Exact** ____________

2. $\begin{array}{r} 3\frac{1}{9} \\ +\ 4\frac{7}{8} \\ \hline \end{array}$

2. **Estimate** ____________ **Exact** ____________

3. $\begin{array}{r} 17\frac{5}{8} \\ 12\frac{1}{4} \\ +\ 5\frac{5}{6} \\ \hline \end{array}$

3. **Estimate** ____________ **Exact** ____________

4. $\begin{array}{r} 126\frac{4}{5} \\ 28\frac{9}{10} \\ +\,13\frac{2}{15} \\ \hline \end{array}$

4. **Estimate**____________

Exact ____________

5. $\begin{array}{r} 26\frac{11}{14} \\ -\,13\frac{5}{18} \\ \hline \end{array}$

5. **Estimate**____________

Exact ____________

6. $\begin{array}{r} 14\frac{4}{7} \\ -\,8\frac{1}{8} \\ \hline \end{array}$

6. **Estimate**____________

Exact ____________

First estimate the answer. Then solve each application problem.

7. A painter used $2\frac{1}{3}$ cans of paint one day and $1\frac{7}{8}$ cans the next day. How many cans did he use altogether?

7. **Estimate**____________

Exact ____________

8. The Eastside Wholesale Vegetable Market sold $4\frac{3}{4}$ tons of broccoli, $8\frac{2}{3}$ tons of spinach, $2\frac{1}{2}$ tons of corn, and $1\frac{5}{12}$ tons of turnips last month. Find the total number of tons of these vegetables sold by the market last month.

8. **Estimate**____________

Exact ____________

9. Paul worked $12\frac{3}{4}$ hours over the weekend. He worked $6\frac{3}{8}$ hours on Saturday. How many hours did he work on Sunday?

9. **Estimate**____________

Exact ____________

Name: Date:
Instructor: Section:

10. On Monday, $7\frac{3}{4}$ tons of cans were recycled, while $9\frac{4}{5}$ tons were recycled on Tuesday. How many more tons were recycled on Tuesday than on Monday?

10.
Estimate ____________
Exact ____________

Objective 2 Estimate an answer, then subtract mixed numbers by regrouping.

First estimate the answer. Then subtract to find the exact answer. Simplify all answers

11. $11\frac{1}{4} - 6\frac{3}{4}$

11.
Estimate ____________
Exact ____________

12. $6\frac{1}{3} - 5\frac{7}{12}$

12.
Estimate ____________
Exact ____________

13. $12\frac{5}{12} - 11\frac{11}{16}$

13.
Estimate ____________
Exact ____________

14. $129\frac{2}{3} - 98\frac{14}{15}$

14.
Estimate ____________
Exact ____________

15. $42 - 19\frac{3}{4}$

15.
Estimate ____________
Exact ____________

16. $\begin{array}{r} 21 \\ -17\frac{9}{16} \\ \hline \end{array}$

16.
Estimate ____________

Exact ____________

17. $\begin{array}{r} 372\frac{5}{6} \\ -208\frac{3}{8} \\ \hline \end{array}$

17.
Estimate ____________

Exact ____________

18. $\begin{array}{r} 147 \\ -39\frac{5}{6} \\ \hline \end{array}$

18.
Estimate ____________

Exact ____________

First estimate the answer. Then solve each application problem.

19. Amy Atwood worked 40 hours during a certain week. She worked $8\frac{1}{4}$ hours on Monday, $6\frac{3}{8}$ hours on Tuesday, $7\frac{3}{4}$ hours on Wednesday, and $8\frac{3}{4}$ hours on Thursday. How many hours did she work on Friday?

19.
Estimate ____________

Exact ____________

20. Three sides of a parking lot are $35\frac{1}{4}$ yards, $42\frac{7}{8}$ yards, and $32\frac{3}{4}$ yards. If the total distance around the lot is $145\frac{1}{2}$ yards, find the length of the fourth side.

20.
Estimate ____________

Exact ____________

Name: Date:
Instructor: Section:

21. A concrete truck is loaded with $11\frac{5}{8}$ cubic yards of concrete. The driver unloads $1\frac{1}{6}$ cubic yards at the first stop, and $2\frac{5}{12}$ cubic yards at the second stop. The customer at the third stop gets 3 cubic yards. How much concrete is left in the truck?

21.
Estimate____________

Exact ____________

22. Debbie Andersen bought 15 yards of material at a sale. She made a shirt with $3\frac{1}{8}$ yards of the material, a dress with $4\frac{7}{8}$ yards, and a jacket with $3\frac{3}{4}$ yards. How many yards of material were left over?

22.
Estimate____________

Exact ____________

Find **x** *in each figure.*

23.

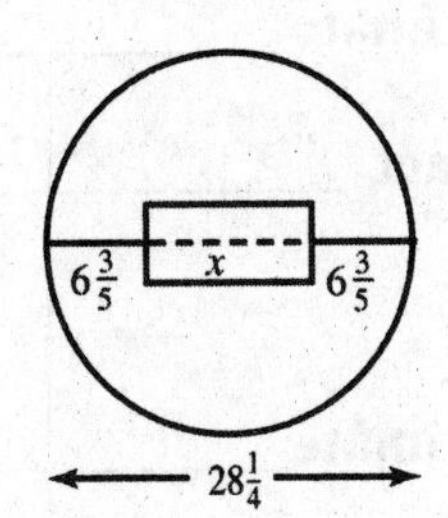

23.
Estimate____________

Exact ____________

24.

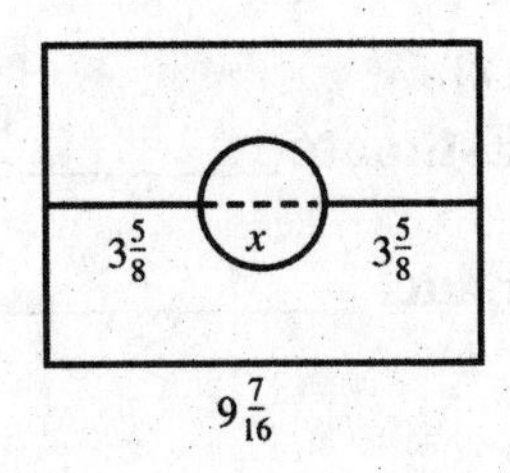

24.
Estimate____________

Exact ____________

Objective 3 Add or subtract mixed numbers using an alternate method.

Add or subtract by changing mixed numbers to improper fractions. Simplify all answers.

25. $5\frac{1}{3}$
$+ 2\frac{5}{6}$

25.
Estimate____________
Exact ____________

26. $1\frac{3}{8}$
$+ 2\frac{3}{5}$

26.
Estimate____________
Exact ____________

27. $3\frac{7}{8}$
$+ 1\frac{5}{12}$

27.
Estimate____________
Exact ____________

28. $3\frac{1}{2}$
$- 1\frac{2}{3}$

28.
Estimate____________
Exact ____________

29. $3\frac{2}{3}$
$- 1\frac{5}{6}$

29.
Estimate____________
Exact ____________

30. $9\frac{1}{8}$
$- 7\frac{4}{9}$

30.
Estimate____________
Exact ____________

Name: Date:
Instructor: Section:

Chapter 3 ADDING AND SUBTRACTING FRACTIONS

3.5 Order Relations and the Order of Operations

Learning Objectives
1 Identify the greater of two fractions.
2 Use exponents with fractions.
3 Use the order of operations with fractions.

Key Terms

Use the vocabulary terms listed below to complete each statement in exercises 1–2.

> <

1. The symbol ____________________ means " is less than."

2. The symbol ____________________ means " is greater than."

Objective 1 Identify the greater of two factions.

Write > or < to make a true statement.

1. $\frac{1}{2}$ ___ $\frac{5}{8}$ 1. ____________

2. $\frac{3}{8}$ ___ $\frac{5}{16}$ 2. ____________

3. $\frac{7}{5}$ ___ $\frac{19}{15}$ 3. ____________

4. $\frac{5}{12}$ ___ $\frac{3}{5}$ 4. ____________

5. $\frac{11}{15}$ ___ $\frac{13}{20}$ 5. ____________

6. $\frac{13}{24}$ ___ $\frac{23}{36}$ 6. ____________

7. $\frac{23}{40}$ ___ $\frac{17}{30}$ 7. ____________

8. $\frac{17}{25}$ ___ $\frac{9}{16}$ 8. ____________

9. $\frac{7}{9}$ ___ $\frac{8}{11}$

9. ______________

Objective 2 Use exponents with fractions.

Simplify. Write the answer in lowest terms.

10. $\left(\frac{1}{4}\right)^3$

10. ______________

11. $\left(\frac{1}{2}\right)^2$

11. ______________

12. $\left(\frac{2}{3}\right)^2$

12. ______________

13. $\left(\frac{5}{3}\right)^3$

13. ______________

14. $\left(\frac{3}{2}\right)^4$

14. ______________

15. $\left(\frac{8}{11}\right)^2$

15. ______________

16. $\left(\frac{1}{2}\right)^4$

16. ______________

17. $\left(\frac{12}{7}\right)^2$

17. ______________

18. $\left(\frac{4}{3}\right)^5$

18. ______________

Name: Date:
Instructor: Section:

Objective 3 Use the order of operations with fractions.

Simplify. Write the answer in lowest terms.

19. $\left(\frac{2}{3}\right)^2 \cdot 6$

19. ______________

20. $\left(\frac{4}{5}\right)^2 \cdot \frac{5}{12}$

20. ______________

21. $\left(\frac{3}{5}\right)^2 \cdot \left(\frac{2}{3}\right)^2$

21. ______________

22. $7 \cdot \left(\frac{2}{7}\right)^2 \cdot \left(\frac{1}{4}\right)^2$

22. ______________

23. $\frac{4}{3} - \frac{1}{2} + \frac{7}{12}$

23. ______________

24. $\frac{1}{2} \cdot \frac{4}{5} + \frac{2}{3} \cdot \frac{9}{5}$

24. ______________

25. $\frac{5}{8} - \frac{2}{3} \cdot \frac{3}{4}$

25. ______________

26. $\frac{3}{4} \cdot \left(\frac{4}{5} + \frac{3}{10}\right)$

26. ______________

27. $\left(\frac{8}{7} - \frac{9}{14}\right) \div \frac{3}{7}$

27. ______________

28. $\frac{8}{7}-\frac{9}{14}\div\frac{9}{7}$

28. ________________

29. $\left(\frac{4}{7}\right)^2\cdot\left(\frac{3}{2}-\frac{5}{8}\right)-\frac{1}{21}\cdot\frac{3}{4}$

29. ________________

30. $\left(\frac{3}{5}\right)^2+\frac{1}{3}\cdot\left(\frac{2}{9}-\frac{1}{5}\right)\div\frac{1}{15}$

30. ________________

Name: Date:
Instructor: Section:

Chapter 4 DECIMALS

4.1 Reading and Writing Decimals

Learning Objectives
1 Write parts of a whole using decimals.
2 Identify the place value of a digit.
3 Read and write decimals in words.
4 Write decimals as fractions or mixed numbers.

Key Terms

Use the vocabulary terms listed below to complete each statement in exercises 1–3.

decimals **decimal point** **place value**

1. We use ____________________ to show parts of a whole.

2. A ____________________ is assigned to each place to the left or right of the decimal point.

3. The dot that separates the whole number part from the fractional part of a decimal number is called the ____________________.

Objective 1 Write parts of a whole using decimals.

Write the portion of each square that is shaded as a fraction, as a decimal, and in words.

1. 1. ____________________
____
____

2. 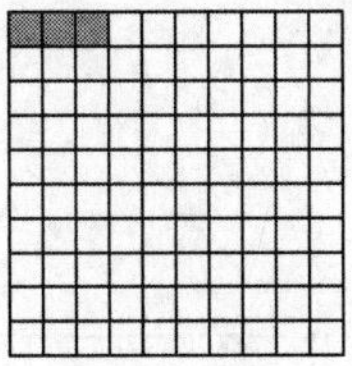 2. ____________________
____
____

3. 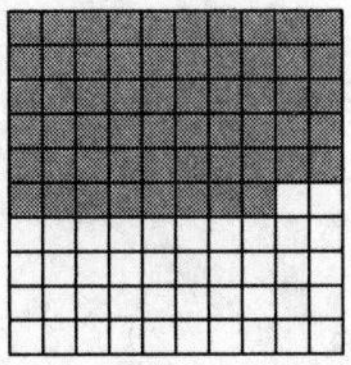 3. ____________________
____
____

Name: Date:
Instructor: Section:

Objective 2 Identify the place value of a digit.

Identify the digit that has the given place value.

4. 43.507 tenths **4.** ________

hundredths ________

5. 0.42583 hundredths **5.** ________

thousandths ________

6. 2.83714 thousandths **6.** ________

ten-thousandths ________

7. 302.9651 hundreds **7.** ________

hundredths ________

Identify the place value of each digit in these decimals.

8. 0.73 7 **8.** ________

3 ________

9. 37.082 3 **9.** ________

7 ________

0 ________

8 ________

2 ________

Objective 3 Read and write decimals in words.

Tell how to read each decimal in words.

10. 0.08 **10.** ________

11. 0.007 **11.** ________

12. 4.06 **12.** ________

Name: Date:
Instructor: Section:

13. 3.0014 **13.** ______________

14. 0.0561 **14.** ______________

15. 10.835 **15.** ______________

16. 2.304 **16.** ______________

17. 97.008 **17.** ______________

Write each decimal in numbers

18. Five and four hundredths **18.** ______________

19. Eleven and nine thousandths **19.** ______________

20. Thirty eight and fifty-two hundred thousandths **20.** ______________

21. Three hundred and twenty-three ten-thousandths **21.** ______________

Use the table below for exercises 22 and 23.

Part Number	Size in Centimeters
7-A	1.08
7-B	1.58
7-C	0.8
8-A	7.02
8-B	7.202

22. Which part number is one and eight hundredths cm? **22.** ______________

23. Write in words the size of part number 8-B. **23.** ______________

Name: Date:
Instructor: Section:

Objective 4 Write decimals as fractions or mixed numbers.

Write each decimal as a fraction or mixed number in lowest terms.

24. 0.8 **24.** ______________

25. 0.001 **25.** ______________

26. 3.6 **26.** ______________

27. 20.0005 **27.** ______________

28. 4.26 **28.** ______________

29. 0.95 **29.** ______________

30. 80.166 **30.** ______________

Name: Date:
Instructor: Section:

Chapter 4 DECIMALS

4.2 Rounding Decimals

Learning Objectives
1 Learn the rules for rounding decimals.
2 Round decimals to any given place.
3 Round money amounts to the nearest cent or nearest dollar.

Key Terms

Use the vocabulary terms listed below to complete each statement in exercises 1–2.

rounding **decimal places**

1. ____________________ are the number of digits to the right of the decimal point.

2. When we "cut off" a number after a certain place value, we are ______________ that number.

Objective 1 Learn the rules for rounding decimals.

Select the phrase that makes the sentence correct.

1. When rounding a number to the nearest tenth, if the digit in the hundredths place is 5 or more, round the digit in the tenths place (up/down). 1. ________________

2. When rounding a number to the nearest hundredth, look at the digit in the (tenth/thousandth) place. 2. ________________

Objective 2 Round decimals to any given place.

Round each number to the place indicated.

3. 17.8937 to the nearest tenth 3. ________________

4. 489.84 to the nearest tenth 4. ________________

5. 785.4982 to the nearest thousandth 5. ________________

6. 43.51499 to the nearest ten-thousandth 6. ________________

7. 54.4029 to the nearest hundredth 7. ________________

Name: Date:
Instructor: Section:

8. 75.399 to the nearest hundredth **8.** ________

9. 989.98982 to the nearest thousandth **9.** ________

10. 486.496 to the nearest one **10.** ________

Round to the nearest hundredth and then to the nearest tenth. Remember to always round the original number.

11. 283.0491 **11.** ________

____

12. 89.525 **12.** ________

____

13. 21.769 **13.** ________

____

14. 0.8948 **14.** ________

____

15. 1.437 **15.** ________

____

16. 0.0986 **16.** ________

____

17. 78.695 **17.** ________

____

18. 108.073 **18.** ________

____

Objective 3 Round money amounts to the nearest cent or nearest dollar.

Round to the nearest dollar.

19. $79.12 **19.** ________

Name: Date:
Instructor: Section:

20. $28.39 **20.** ________________

21. $225.98 **21.** ________________

22. $4797.50 **22.** ________________

23. $11,839.73 **23.** ________________

24. $27,869.57 **24.** ________________

Round to the nearest cent.

25. $1.2499 **25.** ________________

26. $1.0924 **26.** ________________

27. $112.0089 **27.** ________________

28. $134.20506 **28.** ________________

29. $1028.6666 **29.** ________________

30. $2096.0149 **30.** ________________

Name: Date:
Instructor: Section:

Chapter 4 DECIMALS

4.3 Adding and Subtracting Decimals

Learning Objectives
1. Add decimals.
2. Subtract decimals.
3. Estimate the answer when adding or subtracting decimals.

Key Terms

Use the vocabulary terms listed below to complete each statement in exercises 1–2.

estimating **front end rounding**

1. With ______________________________, we round to the highest possible place.

2. Avoid common errors in working decimal problems by ____________________ the answer first.

Objective 1 Add decimals.

Find each sum.

1. 43.96 + 48.53 — 1. ________________

2. 47.94 + 102.38 + 27.631 — 2. ________________

3. 39.87 + 25.2 + 40.36 — 3. ________________

4. 87.6 + 90.4 — 4. ________________

5. 45.83 + 20.923 + 5.7 — 5. ________________

6. 4 + 7.99 + 3.46 — 6. ________________

Name: Date:
Instructor: Section:

7. 10.82 + 5.9 + 4.7 + 6.3 + 20.63 **7.** ______________

Find the perimeter of (distance around) each geometric figure by adding the lengths of the sides.

8. 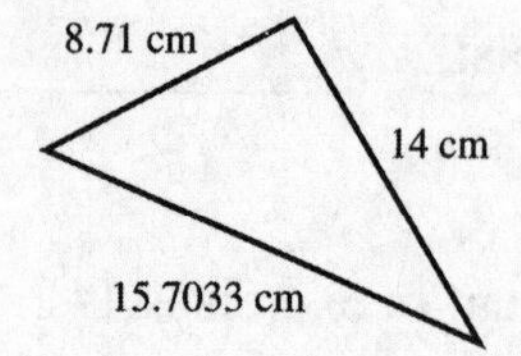

8. ______________

9. 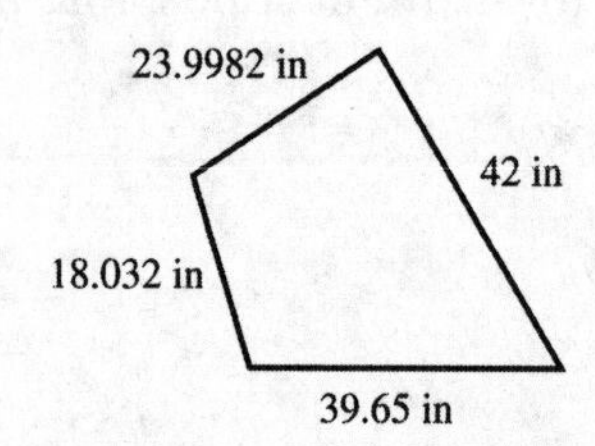

9. ______________

Objective 2 Subtract decimals.

Find each difference.

10. 84.6 – 18.1 **10.** ______________

11. 223.3 – 107.5 **11.** ______________

12. 41.2 – 8.76 **12.** ______________

13. 69.524 – 26.958 **13.** ______________

14. 23.104 – 6.98 **14.** ______________

Name: Date:
Instructor: Section:

15. 71 – 12.68 **15.** ________

16. 689 – 79.832 **16.** ________

Find the unknown measurement in each figure.

17. **17.** ________

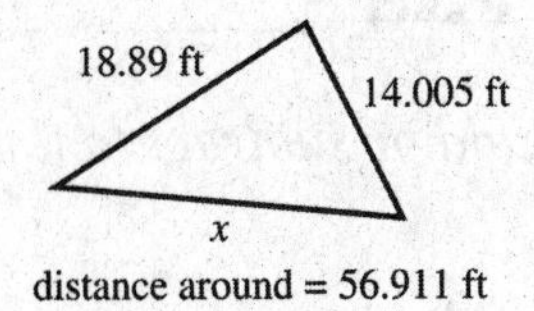

18. **18.** ________

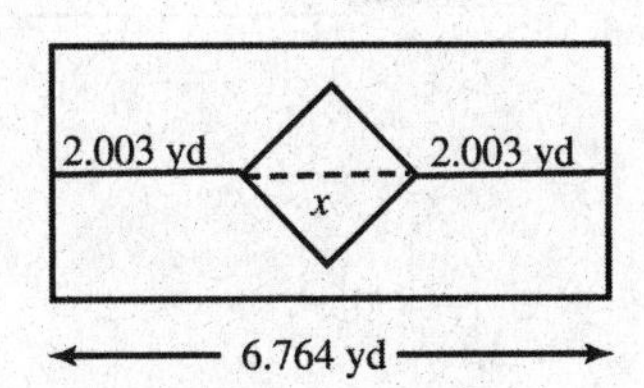

Objective 3 Estimate the answer when adding or subtracting decimals.

First, use front end rounding and estimate each answer. Then add or subtract to find the exact answer.

19.

$$\begin{array}{r} 32.99 \\ 41.72 \\ +\ 8.2 \\ \hline \end{array}$$

19.
Estimate ________
Exact ________

20.

$$\begin{array}{r} 20.85 \\ -\ 7.69 \\ \hline \end{array}$$

20.
Estimate ________
Exact ________

21.

$$\begin{array}{r} 9.7 \\ -\ 4.862 \\ \hline \end{array}$$

21.
Estimate ________
Exact ________

Name: Date:
Instructor: Section:

22.

$$\begin{array}{r} 593.8 \\ 27.93 \\ +\ 54.87 \\ \hline \end{array}$$

22.
Estimate____________

Exact ____________

23.

$$\begin{array}{r} 9 \\ -\ 3.47 \\ \hline \end{array}$$

23.
Estimate____________

Exact ____________

First, use front end rounding and estimate each answer. Then add or subtract to find the exact answer.

24. Kim spent $28.25 for books, $29.47 for a blouse, and $17.85 for a compact disk. How much did she spend?

24.
Estimate____________

Exact ____________

25. Manuel has agreed to work 27.5 hours at a certain job. He has already worked 9.65 hours. How many hours does he have left to work?

25.
Estimate____________

Exact ____________

26. At a fruit stand, Lynn Knight bought $8.53 worth of apples, $11.10 worth of peaches, and $28.29 worth of pears. How much did she spend altogether?

26.
Estimate____________

Exact ____________

27. A customer gives a clerk a $20 bill to pay for $11.29 in purchases. How much change should the customer get?

27.
Estimate____________

Exact ____________

Name: Date:
Instructor: Section:

28. A man receives a bill for $83.26 from Exxon. Of this amount, $53.29 is for a tune-up and the rest is for gas. How much did he pay for gas?

28.
Estimate____________

Exact ____________

29. At the beginning of a trip to El Cerrito, a car odometer read 80,447.5 miles. It is 81.9 miles to El Cerrito. What should the odometer read after driving to El Cerrito and back?

29.
Estimate____________

Exact ____________

30. In India, there are about 50.6 million Internet users, while in Japan, there are about 86.3 million Internet users. How many fewer Internet users are there in India compared to Japan?

30.
Estimate____________

Exact ____________

Chapter 4 DECIMALS

4.4 Multiplying Decimals

Learning Objectives	
1	Multiply decimals.
2	Estimate the answer when multiplying decimals.

Key Terms

Use the vocabulary terms listed below to complete each statement in exercises 1–3.

decimal places **factor** **product**

1. Each number in a multiplication problem is called a ____________________.

2. When multiplying decimal numbers, first find the total number of ______________ in both factors.

3. The answer to a multiplication problem is called the ______________.

Objective 1 Multiply decimals.

Find each product.

1. 0.053×4.3 **1.** ________________

2. 0.682×3.9 **2.** ________________

3. 19.3×4.7 **3.** ________________

4. 96.5×4.6 **4.** ________________

Name: Date:
Instructor: Section:

5. $\begin{array}{r} 67.6 \\ \underline{\times\ 0.023} \end{array}$ **5.** ________

6. $\begin{array}{r} 906 \\ \underline{\times\ 0.081} \end{array}$ **6.** ________

7. (0.074)(0.05) **7.** ________

8. (0.0009)(0.014) **8.** ________

In each of the following, find the amount of money earned on a job by multiplying the number of hours worked and the pay per hour. Round your answer to the nearest cent, if necessary.

9. 27 hours at $6.04 per hour **9.** ________

10. 31.6 hours at $9.83 per hour **10.** ________

Find the cost of each of the following.

11. 16 apples at $0.59 each **11.** ________

12. 7 quarts of oil at $1.05 each **12.** ________

Use the fact that $86 \times 5 = 430$ *to solve exercises 13–18 by simply counting decimal places and writing the decimal point in the correct location.*

13. 86×0.5 **13.** ________

14. 0.86×5 **14.** ________

15. 8.6×0.05 **15.** ______________

16. 0.086×0.05 **16.** ______________

17. 8.6×0.0005 **17.** ______________

18. 0.0086×0.005 **18.** ______________

Solve. Round to the nearest cent, if necessary.

19. The width of Jane's garden is 15.4 feet and the length of the garden is 22.6 feet. What is the area of her garden? **19.** ______________

20. Steve's car payment is $309.56 per month for 48 months. How much will he pay altogether? **20.** ______________

21. The Duncan family's state income tax is found by multiplying the family income of $32,906.15 by the decimal 0.064. Find their tax. **21.** ______________

22. A recycling center pays $0.142 per pound of aluminum. How much would be paid for 176.3 pounds? **22.** ______________

Objective 2 Estimate the answer when multiplying decimals.

First use front-end rounding and estimate the answer. Then multiply to find the exact answer.

23.
$$\begin{array}{r} 49.7 \\ \underline{\times\, 5.8} \end{array}$$

23.
Estimate ______________

Exact ______________

Name: Date:
Instructor: Section:

24. $\begin{array}{r} 29.8 \\ \underline{\times 3.4} \end{array}$

24.
Estimate____________

Exact ____________

25. $\begin{array}{r} 58.73 \\ \underline{\times \ 3.72} \end{array}$

25.
Estimate____________

Exact ____________

26. $\begin{array}{r} 32.53 \\ \underline{\times 23.26} \end{array}$

26.
Estimate____________

Exact ____________

27. $\begin{array}{r} 76.4 \\ \underline{\times 0.57} \end{array}$

27.
Estimate____________

Exact ____________

28. $\begin{array}{r} 2.99 \\ \underline{\times 3.5} \end{array}$

28.
Estimate____________

Exact ____________

29. $\begin{array}{r} 391.9 \\ \underline{\times \ 7.74} \end{array}$

29.
Estimate____________

Exact ____________

30. $\begin{array}{r} 27.5 \\ \underline{\times 11.2} \end{array}$

30.
Estimate____________

Exact ____________

Chapter 4 DECIMALS

4.5 Dividing Decimals

Learning Objectives

1. Divide a decimal by a whole number.
2. Divide a number by a decimal.
3. Estimate the answer when dividing decimals.
4. Use the order of operations with decimals.

Key Terms

Use the vocabulary terms listed below to complete each statement in exercises 1–4.

repeating decimal **quotient** **dividend** **divisor**

1. In a division problem, the number being divided is called the ________________.
2. The number $0.8\bar{3}$ is an example of a ____________________.
3. The answer to a division problem is called the __________________.
4. In the problem $6.39 \div 0.9$, 0.9 is called the __________________.

Objective 1 Divide a decimal by a whole number.

Find each quotient. Round answers to the nearest thousandth, if necessary.

1. $6\overline{)10.763}$ **1.** ________________

2. $5\overline{)34.8}$ **2.** ________________

3. $33\overline{)77.847}$ **3.** ________________

Name: Date:
Instructor: Section:

4. $11\overline{)46.98}$ **4.** ____________

5. $54\overline{)895.79}$ **5.** ____________

Solve. Round to the nearest cent, if necessary.

6. To build a barbecue, Diana Jenkins bought 589 bricks, paying $185.70. Find the cost per brick. **6.** ____________

Objective 2 Divide a number by a decimal.

Find each quotient. Round answers to the nearest thousandth, if necessary.

7. $0.9\overline{)3.4166}$ **7.** ____________

8. $3.4\overline{)436.05}$ **8.** ____________

9. $2859.4 \div 0.053$ **9.** ____________

10. $0.07 \div 0.00043$ **10.** ____________

Solve each application problem. Round money answers to the nearest cent, if necessary.

11. Leon Williams drove 542.2 miles on the 16.3 gallons of gas in his Ford Taurus. How many miles per gallon did he get? Round to the nearest tenth.

11. ________________

12. Lakesha Starr bought 7.4 yards of fabric, paying a total of $26.27. Find the cost per yard.

12. ________________

Objective 3 Estimate the answer when dividing decimals.

Decide if each answer is ***reasonable*** *or* ***unreasonable*** *by rounding the numbers and estimating the answer.*

13. $49.8 \div 7.1 = 7.014$

13. ________________

14. $126.2 \div 11.2 = 11.268$

14. ________________

15. $31.5 \div 8.4 = 37.5$

15. ________________

16. $486.9 \div 5.06 = 962.253$

16. ________________

17. $1092.8 \div 37.92 = 2.882$

17. ________________

18. $1564.9 \div 50.049 = 312.674$

18. ________________

19. $8695.15 \div 98.762 = 88.0415$

19. ________________

20. $6608.04 \div 415.6 = 15.9$

20. ________________

Name: Date:
Instructor: Section:

Objective 4 Use the order of operations with decimals.

Use the order of operations to simplify each expression.

21. $3.7 + 5.1^2 - 9.4$ **21.** ____________

22. $3.1^2 - 1.9 + 5.8$ **22.** ____________

23. $42.92 \div 5.8 \times 7.3$ **23.** ____________

24. $55.744 \div (6.4 \times 1.9)$ **24.** ____________

25. $18.5 + (37.1 - 29.8)(10.7)$ **25.** ____________

26. $58.1 - (17.9 - 15.2) \times 1.8$ **26.** ____________

27. $27.51 - 3.2 \times 9.8 \div 1.6$ **27.** ____________

28. $9.1 - 0.07(2.1 \div 0.042)$ **28.** ____________

29. $9.8 \times 4.76 + 17.94 \div 2.6$ **29.** ____________

30. $62.699 \div 7.42 + 3.6 \times 1.4$ **30.** ____________

Chapter 4 DECIMALS

4.6 Writing Fractions as Decimals

Learning Objectives
1 Write fractions as equivalent decimals.
2 Compare the size of fractions and decimals.

Key Terms

Use the vocabulary terms listed below to complete each statement in exercises 1–4.

numerator **denominator** **mixed number** **equivalent**

1. A fraction and a decimal that represent the same portion of a whole are ________________.

2. The ____________________ of a fraction is the dividend.

3. The __________________ of a fraction shows the number of equal parts in a whole.

4. A __________________ consists of a whole number part and a fractional or decimal part.

Objective 1 Write fractions as equivalent decimals.

Write each fraction or mixed number as a decimal. Round to the nearest thousandth, if necessary.

1. $6\frac{1}{2}$ 1. ________________

2. $\frac{1}{5}$ 2. ________________

3. $2\frac{2}{3}$ 3. ________________

4. $\frac{1}{8}$ 4. ________________

5. $\frac{1}{11}$ 5. ________________

6. $7\frac{1}{10}$ 6. ________________

Name: Date:
Instructor: Section:

7. $\frac{3}{5}$ 7. ____________

8. $\frac{7}{8}$ 8. ____________

9. $4\frac{1}{9}$ 9. ____________

10. $\frac{13}{25}$ 10. ____________

11. $\frac{3}{20}$ 11. ____________

12. $31\frac{3}{13}$ 12. ____________

13. $19\frac{17}{24}$ 13. ____________

14. Jose got 4 hits in 11 times at bat. Write his batting average as a decimal rounded to the nearest thousandth. 14. ____________

Objective 2 Compare the size of fractions and decimals.

Write $<$ or $>$ to make a true statement.

15. $\frac{5}{8}$ ___ 0.634 15. ____________

16. $\frac{5}{6}$ ___ 0.83 16. ____________

17. $\frac{1}{25}$ ___ 0.039 17. ____________

18. $\frac{3}{16}$ ___ 0.188 18. ____________

19. $\frac{5}{9}$ ___ 0.55 19. ____________

20. $\frac{3}{8}$ ___ 0.38 **20.** ________________

21. $0.\overline{7}$ ____ 0.7 **21.** ________________

22. Candy bar A weighs 1.4 ounces, while candy bar B weighs $1\frac{3}{8}$ ounces. Which weighs more? **22.** ________________

Arrange in order from smallest to largest.

23. $\frac{7}{15}$, 0.466, $\frac{9}{19}$ **23.** ________________

24. $\frac{8}{9}$, 0.88, 0.89 **24.** ________________

25. $\frac{3}{11}$, $\frac{1}{3}$, 0.29 **25.** ________________

26. $\frac{1}{7}$, $\frac{3}{16}$, 0.187 **26.** ________________

27. 0.8462, $\frac{11}{13}$, $\frac{6}{7}$ **27.** ________________

28. 1.085,$1\frac{5}{11}$,$1\frac{7}{20}$ **28.** ________________

29. 0.16666,$\frac{1}{6}$,0.1666,0.01666 **29.** ________________

30. Five cyclists in a race had the following times: **30.** ________________
Catherine, 11.06 min Edita, 10.53 min
Olga, 11.24 min Diana, 10.51
Anna, 10.38 min
List them in the order they placed.

Name: Date:
Instructor: Section:

Chapter 5 RATIO AND PROPORTION

5.1 Ratios

Learning Objectives

1. Write ratios as fractions.
2. Solve ratio problems involving decimals or mixed numbers.
3. Solve ratio problems after converting units.

Key Terms

Use the vocabulary terms listed below to complete each statement in exercises 1–2.

denominator **numerator** **ratio**

1. A ____________________ can be used to compare two measurements with the same type of units.

2. When writing the ratio to compare the width of a room to its height, the width goes in the ____________________ and the height goes in the ____________________.

Objective 1 Write ratios as fractions.

Write each ratio as a fraction in lowest terms.

1. 18 to 24 — 1. ____________

2. 76 to 101 — 2. ____________

3. 125 cents to 95 cents — 3. ____________

4. 80 miles to 30 miles — 4. ____________

5. $85 to $135 — 5. ____________

6. 5 men to 20 men — 6. ____________

Solve. Write each ratio as a fraction in lowest terms.

7. Mr. Williams is 42 years old, and his son is 18. Find the ratio of Mr. Williams' age to his son's age. — 7. ____________

Name: Date:
Instructor: Section:

8. When using Roundup vegetation control, add 128 ounces of water for every 6 ounces of the herbicide. Find the ratio of herbicide to water. **8.** ________

Objective 2 Solve ratio problems involving decimals or mixed numbers.

Write each ratio as a fraction in lowest terms.

9. $6\frac{1}{2}$ to 2 **9.** ________

10. $4\frac{1}{8}$ to 3 **10.** ________

11. 3 to $2\frac{1}{2}$ **11.** ________

12. 11 to $2\frac{4}{9}$ **12.** ________

13. $1\frac{1}{4}$ to $1\frac{1}{2}$ **13.** ________

14. $3\frac{1}{2}$ to $1\frac{3}{4}$ **14.** ________

Solve. Write each ratio as a fraction in lowest terms.

15. One refrigerator holds $3\frac{3}{4}$ cubic feet of food, while another holds 5 cubic feet. Find the ratio of the amount of storage in the first refrigerator to the amount of storage in the second. **15.** ________

16. One car has a $15\frac{1}{2}$ gallon gas tank while another has a 22 gallon gas tank. Find the ratio of the amount the first tank holds to the amount the second tank holds. **16.** ________

17. The price of gasoline increased from $2.75 per gallon to $3.25 per gallon. Find the ratio of the increase in price to the original price. **17.** ________

Name: Date:
Instructor: Section:

18. The amount of cereal in a giant-size box decreased from 27.6 ounces to 22.4 ounces. Find the ratio of the original size to the new size.

18. ______________

For each triangle, find the ratio of the length of the longest side to the length of the shortest side. Write each ratio as a fraction in lowest terms.

19.

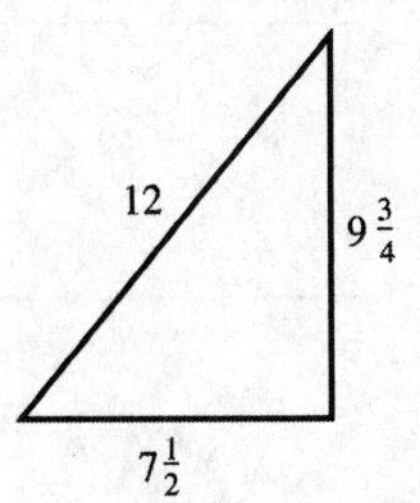

19. ______________

20.

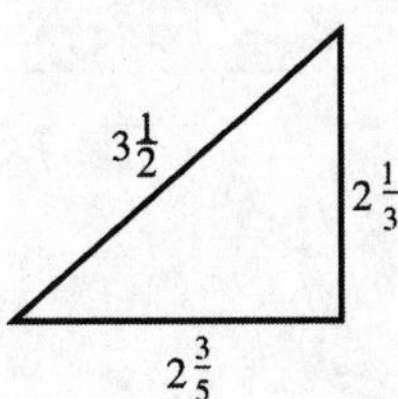

20. ______________

Objective 3 Solve ratio problems after converting units.

Write each ratio as a fraction in lowest terms. Be sure to convert units as necessary.

21. 4 days to 2 weeks

21. ______________

22. 4 feet to 15 inches

22. ______________

23. 6 yards to 10 feet

23. ______________

24. 7 gallons to 8 quarts

24. ______________

25. 40 ounces to 3 pounds

25. ______________

Name: Date:
Instructor: Section:

26. 80 cents to $3 **26.** ______________

27. Find the ratio of $17\frac{1}{2}$ inches to $2\frac{1}{3}$ feet. **27.** ______________

28. What is the ratio of $59\frac{1}{2}$ ounces to $4\frac{1}{4}$ pounds? **28.** ______________

29. What is the ratio of $9\frac{1}{3}$ yards to $3\frac{1}{2}$ feet? **29.** ______________

30. A tree is $28\frac{3}{4}$ feet tall. It casts a shadow 81 inches long. Find the ratio of the height of the tree to the length of its shadow. **30.** ______________

Name: Date:
Instructor: Section:

Chapter 5 RATIO AND PROPORTION

5.2 Rates

Learning Objectives	
1	Write rates as fractions.
2	Find unit rates.
3	Find the best buy based on cost per unit.

Key Terms

Use the vocabulary terms listed below to complete each statement in exercises 1–3.

rate **unit rate** **cost per unit**

1. When the denominator of a rate is 1, it is called a ______________________.

2. The ______________________ is that rate that tells how much is paid for one item.

3. A ____________________ compares two measurements with different units.

Objective 1 Write rates as fractions.

Write each rate as a fraction in lowest terms.

1. 75 miles in 25 minutes **1.** ________________

2. 85 feet in 17 seconds **2.** ________________

3. 28 dresses for 4 women **3.** ________________

4. 70 horses for 14 teams **4.** ________________

5. 45 gallons in 3 hours **5.** ________________

6. 225 miles on 15 gallons **6.** ________________

7. 119 pills for 17 patients **7.** ________________

8. 144 kilometers on 16 liters **8.** ________________

9. 256 pages for 8 chapters **9.** ________

10. 990 miles in 18 hours **10.** ________

Objective 2 Find unit rates.

Find each unit rate.

11. \$75 in 5 hours **11.** ________

12. \$3500 in 20 days **12.** ________

13. \$1540 in 14 days **13.** ________

14. \$7875 for 35 pounds **14.** ________

15. \$122.76 in 9 hours **15.** ________

16. 189.88 miles on 9.4 gallons **16.** ________

Solve each application problem.

17. Eric can pack 12 crates of berries in 24 minutes. Give his rate in crates per minute and in minutes per crate. **17.** ________

____

18. Michelle can plow 7 acres in 14 hours. Give her rate in acres per hour and in hours per acre. **18.** ________

____

19. Meili earns \$220.32 in 24 hours. What is her rate per hour? **19.** ________

Name: Date:
Instructor: Section:

20. The 4.6 yards of fabric needed for a dress costs $27.14. Find the cost of 1 yard. **20.** ______________

21. A company pays $3225 in dividends for the 1250 shares of its stock. Find the value of dividends per share. **21.** ______________

22. Jojo drove 434 miles on 15.5 gallons of gasoline. How many miles did he drive per gallon? **22.** ______________

23. It took Marlene $7\frac{1}{4}$ hours to drive 450 miles. What was her average speed per hour? **23.** ______________

Objective 3 Find the best buy based on cost per unit.

Find the best buy (based on cost per unit) for each item.

24. Beans: 12 ounces for $1.49; 16 ounces of $1.89 **24.** ______________

25. Orange juice: 16 ounces for $0.89; 32 ounces for $1.90 **25.** ______________

26. Peanut butter: 18 ounces for $1.77; 24 ounces for $2.08 **26.** ______________

27. Batteries: 4 for $2.79; 10 for $4.19 **27.** ______________

Name: Date:
Instructor: Section:

28. Cola: 6 cans for \$1.98; 12 cans for \$3.59; 24 cans for \$8

28. ________________

29. Soup: 3 cans for \$1.75; 5 cans for \$2.75; 8 cans for \$4.55

29. ________________

30. Cereal: 10 ounces for \$1.34; 15 ounces for \$1.76; 20 ounces for \$2.29

30. ________________

Name: Date:
Instructor: Section:

Chapter 5 RATIO AND PROPORTION

5.3 Proportions

Learning Objectives
1. Write proportions.
2. Determine whether proportions are true or false.
3. Find cross products.

Key Terms

Use the vocabulary terms listed below to complete each statement in exercises 1–2.

cross products **proportion**

1. A ______________________________ shows that two ratios or rates are equivalent.
2. To see whether a proportion is true, determine if the ____________________ are equal.

Objective 1 Write proportions.

Write each proportion.

1. 11 is to 15 as 22 is to 30. 1. ________________
2. 50 is to 8 as 75 is to 12. 2. ________________
3. 24 is to 30 as 8 is to 10. 3. ________________
4. 36 is to 45 as 8 is to 10. 4. ________________
5. 14 is to 21 as 10 is to 15. 5. ________________
6. 3 is to 33 as 12 is to 132. 6. ________________

Name: Date:
Instructor: Section:

7. $1\frac{1}{2}$ is to 4 as 21 is to 56. **7.** ______________

8. $3\frac{2}{3}$ is to 11 as 10 is to 30. **8.** ______________

9. $6\frac{2}{5}$ is to 12 as 8 is to 3. **9.** ______________

Objective 2 Determine whether proportions are true or false.

Determine whether each proportion is true or false by writing the ratios in lowest terms. Show the simplified ratios and then write **true** *or* **false**.

10. $\frac{6}{100} = \frac{3}{50}$ **10.** ______________

11. $\frac{48}{36} = \frac{3}{4}$ **11.** ______________

12. $\frac{3}{8} = \frac{21}{28}$ **12.** ______________

13. $\frac{30}{25} = \frac{6}{5}$ **13.** ______________

14. $\frac{390}{100} = 27$ **14.** ______________

15. $\frac{35}{21} = \frac{3}{4}$ **15.** ______________

Name: Date:
Instructor: Section:

16. $\dfrac{28}{6}=\dfrac{42}{9}$

16. ______________

17. $\dfrac{54}{30}=\dfrac{108}{60}$

17. ______________

18. $\dfrac{15}{24}=\dfrac{25}{35}$

18. ______________

19. $\dfrac{63}{18}=\dfrac{56}{14}$

19. ______________

20. $\dfrac{108}{225}=\dfrac{24}{50}$

20. ______________

Objective 3 Find cross products.

Use cross products to determine whether each proportion is true or false. Show the cross products and then write **true** *or* **false**.

21. $\dfrac{10}{45}=\dfrac{6}{27}$

21. ______________

22. $\dfrac{28}{50}=\dfrac{49}{75}$

22. ______________

23. $\dfrac{132}{24}=\dfrac{11}{3}$

23. ______________

Name: Date:
Instructor: Section:

24. $\dfrac{3\frac{1}{2}}{4} = \dfrac{14}{16}$ **24.** ______________

25. $\dfrac{4\frac{3}{5}}{9} = \dfrac{18\frac{2}{5}}{36}$ **25.** ______________

26. $\dfrac{21}{28} = \dfrac{5\frac{3}{4}}{7}$ **26.** ______________

27. $\dfrac{22}{54} = \dfrac{6\frac{1}{3}}{5\frac{2}{11}}$ **27.** ______________

28. $\dfrac{69.9}{3} = \dfrac{100.19}{4.3}$ **28.** ______________

29. $\dfrac{2.98}{7.1} = \dfrac{1.7}{4.3}$ **29.** ______________

30. $\dfrac{42.2}{106.8} = \dfrac{84.9}{206}$ **30.** ______________

Name: Date:
Instructor: Section:

Chapter 5 RATIO AND PROPORTION

5.4 Solving Proportions

Learning Objectives
1. Find the unknown number in a proportion.
2. Find the unknown number in a proportion with mixed numbers or decimals.

Key Terms

Use the vocabulary terms listed below to complete each statement in exercises 1–3.

cross products **proportion** **ratio**

1. A ____________________ is a statement that two ratios are equal.
2. The ____________________ of the proportion $\frac{a}{b} = \frac{c}{d}$ are *ad* and *bc*.
3. A ____________________ is a comparison of two quantities with the same units.

Objective 1 Find the unknown number in a proportion.

Find the unknown number in each proportion.

1. $\frac{3}{2} = \frac{x}{6}$ 1. ____________

2. $\frac{9}{4} = \frac{36}{x}$ 2. ____________

3. $\frac{9}{7} = \frac{x}{28}$ 3. ____________

4. $\frac{x}{11} = \frac{44}{121}$ 4. ____________

5. $\frac{35}{x} = \frac{5}{3}$ 5. ____________

Name: Date:
Instructor: Section:

6. $\dfrac{x}{52}=\dfrac{5}{13}$

6. ________________

7. $\dfrac{96}{60}=\dfrac{8}{x}$

7. ________________

8. $\dfrac{7}{5}=\dfrac{98}{x}$

8. ________________

9. $\dfrac{9}{14}=\dfrac{x}{70}$

9. ________________

10. $\dfrac{90}{x}=\dfrac{15}{8}$

10. ________________

11. $\dfrac{x}{110}=\dfrac{7}{10}$

11. ________________

12. $\dfrac{14}{x}=\dfrac{21}{18}$

12. ________________

13. $\dfrac{18}{81}=\dfrac{4}{x}$

13. ________________

14. $\dfrac{100}{x}=\dfrac{75}{30}$

14. ________________

Name: Date:
Instructor: Section:

15. $\frac{x}{45} = \frac{132}{180}$ 15. ________________

Objective 2 Find the unknown number in a proportion with mixed numbers or decimals.

Find the unknown number in each proportion. Write answers as a whole or a mixed number if possible.

16. $\frac{2}{3\frac{1}{4}} = \frac{8}{x}$ 16. ________________

17. $\frac{3}{x} = \frac{5}{1\frac{2}{3}}$ 17. ________________

18. $\frac{x}{6} = \frac{5\frac{1}{4}}{7}$ 18. ________________

19. $\frac{1\frac{1}{5}}{\frac{1}{2}} = \frac{6}{x}$ 19. ________________

20. $\frac{0}{5\frac{1}{3}} = \frac{x}{5}$ 20. ________________

21. $\frac{x}{7\frac{1}{2}} = \frac{3}{6\frac{2}{3}}$ 21. ________________

22. $\frac{3}{x} = \frac{0.8}{5.6}$ 22. ________________

Name: Date:
Instructor: Section:

23. $\frac{16}{12}=\frac{2}{x}$

23. ________________

24. $\frac{4.2}{x}=\frac{0.6}{2}$

24. ________________

25. $\frac{2\frac{1}{2}}{1\frac{2}{3}}=\frac{x}{2}$

25. ________________

26. $\frac{2\frac{5}{9}}{x}=\frac{23}{\frac{3}{5}}$

26. ________________

27. $\frac{10}{x}=\frac{2\frac{1}{2}}{2}$

27. ________________

28. $\frac{x}{7.9}=\frac{0}{47.4}$

28. ________________

29. $\frac{x}{4.8}=\frac{1.5}{1.2}$

29. ________________

30. $\frac{32}{2.4}=\frac{x}{3}$

30. ________________

Name: Date:
Instructor: Section:

Chapter 5 RATIO AND PROPORTION

5.5 Solving Application Problems with Proportions

Learning Objectives
1 Use proportions to solve application problems.

Key Terms

Use the vocabulary terms listed below to complete each statement in exercises 1–2.

rate **ratio**

1. A statement that compares a number of inches to a number of inches is a ________________.

2. A statement that compares a number of gallons to a number of miles is a ____________________.

Objective 1 Use proportions to solve application problems.

Set up and solve a proportion for each problem.

1. A gardening service charges $45 to install 50 square feet of sod. Find the charge to install 125 feet. **1.** ________________

2. On a road map, a length of 3 inches represents a distance of 8 miles. How many inches represent a distance of 32 miles? **2.** ________________

3. If 6 melons cost $9, find the cost of 10 melons. **3.** ________________

4. If 22 hats cost $198, find the cost of 12 hats. **4.** ________________

Name: Date:
Instructor: Section:

5. 6 pounds of grass seed cover 4200 square feet of ground. How many pounds are needed for 5600 square feet.

5. ________________

6. Margie earns $168.48 in 26 hours. How much does she earn in 40 hours?

6. ________________

7. Juan makes $477.40 in 35 hours. How much does he make in 60 hours.

7. ________________

8. If 5 ounces of a medicine must be mixed with 12 ounces of water, how many ounces of medicine would be mixed with 132 ounces of water.

8. ________________

9. The distance between two cities on a road map is 5 inches. The two cities are really 600 miles apart. The distance between two other cities on the map is 8 inches. How many miles apart are these cities?

9. ________________

10. The distance between two cities is 600 miles. On a map the cities are 10 inches apart. Two other cities are 720 miles apart. How many inches apart are they on the map?

10. ________________

Name: Date:
Instructor: Section:

11. If 2 visits to a salon cost \$80, find the cost of 11 visits. **11.** ________________

12. If a 4-minute phone call costs \$0.96, find the cost of a 10-minute call. **12.** ________________

13. If 150 square yards of carpet cost \$3142.50, find the cost of 210 square yards of the carpet. **13.** ________________

14. Scott paid \$240,000 for a 5-unit apartment house. Find the cost of a 16-unit apartment house. **14.** ________________

15. Brian plants his seeds early in the year. To keep them from freezing, he covers the ground with black plastic. A piece with an area of 80 square feet costs \$14. Find the cost of a piece with an area of 700 square feet. **15.** ________________

16. A taxi ride of 7 miles costs \$9.45. Find the cost of a ride of 12 miles. **16.** ________________

17. Dog food for 8 dogs costs \$15. Find the cost of dog food for 12 dogs. **17.** ________________

Name: Date:
Instructor: Section:

18. To make battery acid, Jeff mixes $9\frac{1}{2}$ gallons of pure acid with 25 gallons of water. How much acid would be needed for 75 gallons of water?

18. ______________

19. Tax on an $18,000 car is $1620. Find the tax on a $24,000 car.

19. ______________

20. If $18\frac{3}{4}$ yards of material are needed for 5 dresses, how much material is needed for 9 dresses?

20. ______________

21. If it takes 6 minutes to read 4 pages of a book, how long will it take to read 320 pages?

21. ______________

22. If a gallon of paint will cover 400 square feet, how many gallons are needed to cover 2200 square feet?

22. ______________

23. The height of the water in a fish tank rises at a constant rate of 2 inches every 5 minutes. How many minutes will it take to fill the tank if the height must reach 25 inches?

23. ______________

24. It costs $15 dollars to park for 4 hours. How long will you have parked a car if your cost is $25 dollars?

24. ______________

Name: Date:
Instructor: Section:

25. A ball that is dropped from a height of 60 inches will rebound to a height of 48 inches. How high will a ball rebound that is dropped from a height of 96 inches?

25. _______________

26. A person weighing 150 pounds on Earth weighs approximately 25 pounds on the moon. How much will a person weigh on Earth if their moon weight is 32 pounds?

26. _______________

27. Five apples cost $1.60. How much will 8 apples cost?

27. _______________

28. A biologist tags 50 deer and releases them in a wildlife preserve area. Over the course of a two-week period, she observes 80 deer, of which 12 are tagged. What is the estimate for the population of deer in this particular area?

28. _______________

29. A paving crew completes 10,000 feet of a road every 3 days. Approximately how many days will it take to pave a 7-mile stretch of road? (1 mile = 5280 feet)

29. _______________

30. A model airplane has a wingspan of 8 inches. The actual wingspan of the plane it represents is 38 feet. If the model's fuselage is 12 inches long, how long is the fuselage of the actual plane?

30. _______________

Name: Date:
Instructor: Section:

Chapter 6 PERCENT

6.1 Basics of Percent

Learning Objectives

1. Learn the meaning of percent.
2. Write percents as decimals.
3. Write decimals as percents.
4. Understand 100%, 200%, and 300%.
5. Use 50%, 10%, and 1%.

Key Terms

Use the vocabulary terms listed below to complete each statement in exercises 1–3.

percent **ratio** **decimals**

1. To compare two quantities that have the same type of units, use a ____________.

2. ____________________ means per one hundred.

3. ____________________ represent parts of a whole.

Objective 1 Learn the meaning of percent.

Write as a percent.

1. 43 people out of 100 drive small cars. **1.** ________________

2. The tax is $8 per $100. **2.** ________________

3. The cost for labor was $45 for every $100 spent to manufacture an item. **3.** ________________

4. 38 out of 100 planes departed on time. **4.** ________________

Objective 2 Write percents as decimals.

Write each percent as a decimal.

5. 42% **5.** ________________

6. 310% **6.** ________________

7. 4% **7.** ________________

8. 10% **8.** ________________

Name: Date:
Instructor: Section:

9. 2.5% **9.** ____________

10. 0.025% **10.** ____________

11. 0.256% **11.** ____________

Objective 3 Write decimals as percents.

Write each decimal as a percent.

12. 0.30 **12.** ____________

13. 0.2 **13.** ____________

14. 0.07 **14.** ____________

15. 0.564 **15.** ____________

16. 4.93 **16.** ____________

17. 5.5 **17.** ____________

18. 0.036 **18.** ____________

Objective 4 Understand 100%, 200%, and 300%.

Fill in the blanks.

19. 100% of $19 is __________. **19.** ____________

20. 200% of 170 miles is __________. **20.** ____________

21. 300% of $76 is __________. **21.** ____________

Name: Date:
Instructor: Section:

22. 100% of 12 dogs is __________. **22.** __________________

23. 200% of $520 is __________. **23.** __________________

24. 300% of $250 is __________. **24.** __________________

Objective 5 Use 50%, 10%, and 1%.

Fill in the blanks.

25. 50% of 250 signs is __________ **25.** __________________

26. 10% of 100 years is __________. **26.** __________________

27. 50% of 48 copies is __________. **27.** __________________

28. 10% of 4920 televisions is __________. **28.** __________________

29. 1% of 400 homes is __________. **29.** __________________

30. 1% of $98 is __________. **30.** __________________

Name: Date:
Instructor: Section:

Chapter 6 PERCENT

6.2 Percents and Fractions

Learning Objectives
1 Write percents as fractions.
2 Write fractions as percents.
3 Use the table of percent equivalents.

Key Terms

Use the vocabulary terms listed below to complete each statement in exercises 1–2.

percent **lowest terms**

1. A fraction is in ________________ when its numerator and denominator have no common factor other than 1.

2. A ________________ can be written as a fraction with 100 in the denominator.

Objective 1 Write percents as fractions.

Write each percent as a fraction or mixed number in lowest terms.

1. 12% 1. ____________

2. 86% 2. ____________

3. 62.5% 3. ____________

4. 43.6% 4. ____________

5. $16\frac{2}{3}\%$ 5. ____________

6. $22\frac{2}{9}\%$ 6. ____________

7. 0.5% 7. ____________

8. 0.04% 8. ____________

Name: Date:
Instructor: Section:

9. 140% **9.** ______

10. 275% **10.** ______

Objective 2 Write fractions as percents.

Write each fraction or mixed number as a percent. Round percents to the nearest tenth, if necessary.

11. $\frac{7}{10}$ **11.** ______

12. $\frac{81}{100}$ **12.** ______

13. $\frac{12}{25}$ **13.** ______

14. $\frac{64}{75}$ **14.** ______

15. $\frac{47}{50}$ **15.** ______

16. $\frac{5}{9}$ **16.** ______

17. $3\frac{4}{5}$ **17.** ______

Name: Date:
Instructor: Section:

18. $2\frac{3}{4}$ **18.** ________________

19. $7\frac{2}{5}$ **19.** ________________

20. $4\frac{1}{3}$ **20.** ________________

Objective 3 Use the table of percent equivalents.

Complete this chart. Round decimals to the nearest thousandth and percents to the nearest tenth, if necessary.

	Fraction	*Decimal*	*Percent*	
21.	$\frac{1}{2}$	______	______	**21.** ________________
22.	______	0.125	______	**22.** ________________
23.	$\frac{1}{4}$	______	______	**23.** ________________
24.	$\frac{5}{8}$	______	______	**24.** ________________
25.	______	______	87.5%	**25.** ________________

Name: Date:
Instructor: Section:

26.	$\frac{3}{8}$	______	______	**26.** ____________
27.	______	______	$33\frac{1}{3}\%$	**27.** ____________
28.	$\frac{2}{5}$	______	______	**28.** ____________
29.	______	0.325	______	**29.** ____________
30.	$\frac{2}{3}$	______	______	**30.** ____________

Name: Date:
Instructor: Section:

Chapter 6 PERCENT

6.3 Using the Percent Proportion and Identifying the Components in a Percent Problem

Learning Objectives

1. Learn the percent proportion.
2. Solve for an unknown value in a percent proportion.
3. Identify the percent.
4. Identify the whole.
5. Identify the part.

Key Terms

Use the vocabulary terms listed below to complete each statement in exercises 1–3.

percent proportion **whole** **part**

1. The ______________________________ in a percent problem is the entire quantity.

2. The ____________________ in a percent problem is the portion being compared with the whole.

3. Part is to whole as percent is to 100 is called the ________________________.

Objective 1 Learn the percent proportion.

1. Write the percent proportion **1.** ________________

Objective 2 Solve for an unknown value in a percent proportion.

Use the percent proportion to solve for the unknown value. Round to the nearest tenth, if necessary. If the answer is a percent, be sure to include a percent sign.

2. part = 30, percent = 25 **2.** ________________

3. part = 160, percent = 20 **3.** ________________

4. part = 18, percent = 150 **4.** ________________

5. whole = 48, percent = 25 **5.** ________________

Name: Date:
Instructor: Section:

6. whole = 25, percent = 14 **6.** ______________

7. whole = 50, percent = 175 **7.** ______________

8. part = 12, whole = 50 **8.** ______________

9. part = 75, whole = 1500 **9.** ______________

10. part = 160, whole = 120 **10.** ______________

Objective 3 Identify the percent.

Identify the percent. Do not try to solve for any unknowns.

11. 83% of what number is 21.5? **11.** ______________

12. 36 is 72% of what number? **12.** ______________

Identify the percent in each application problem. Do not try to solve for any unknowns.

13. A chemical is 42% pure. Of 800 grams of the chemical, how much is pure? **13.** ______________

14. Sales tax of $8 is charged on an item costing $200. What percent of sales tax is charged? **14.** ______________

15. 17% of Tom's check of $340 is withheld. How much is withheld? **15.** ______________

Name: Date:
Instructor: Section:

16. A team won 12 of the 18 games it played. What percent of its games did it win? **16.** ______________

Objective 4 Identify the whole.

Identify the whole. Do not try to solve for any unknowns.

17. 71 is what percent of 384? **17.** ______________

18. 0.68% of 487 is what number? **18.** ______________

19. What is 14% of 78? **19.** ______________

Identify the whole in each application problem. Do not try to solve for any unknowns.

20. In one storm, Springbrook got 15% of the season's snowfall. Springbrook's total snowfall for that season was 30 inches. How many inches of snow fell in that one storm? **20.** ______________

21. In one state, the sales tax is 8%. On a purchase, the amount of tax was $26. Find the cost of the item purchase. **21.** ______________

22. In an election, 68% of the registered voters actually voted. If there are 12,452 voters, how many people voted? **22.** ______________

Objective 5 Identify the part.

Identify the part. Do not try to solve for any unknowns.

23. 29.81 is what percent of 508? **23.** ______________

Name: Date:
Instructor: Section:

24. 16.74 is 11.9% of what number? **24.** ______________

25. What number is 12.4% of 1408? **25.** ______________

Identify the part, then set up the percent proportion in each application problem. Do not try to solve for any unknowns.

26. In a one-day storm, Odentown received 0.3% of the season's total rainfall. Odentown received 4 inches of rain on that day. How many inches of rain fell during the season? **26.** ______________

27. A hatchery is notified that 7% of its shipment of baby salmon did not arrive healthy. Of 1500 salmon shipped, how many did not arrive healthy? **27.** ______________

28. There are 720 quarts of grape juice in a vat holding a total of 2400 quarts of fruit juice. What percent of the vat is grape juice? **28.** ______________

29. A teacher of English literature found that 15% of the students' papers are handed in late. If there are 40 students in a class, how many papers will be handed in late? **29.** ______________

30. Payroll deductions are 35% of Jason's gross pay. If his deductions total $350, what is his gross pay? **30.** ______________

Name: Date:
Instructor: Section:

Chapter 6 PERCENT

6.4 Using Proportions to Solve Percent Problems

Learning Objectives	
1	Use the percent proportion to find the part.
2	Find the whole using the percent proportion.
3	Find the percent using the percent proportion.

Key Terms

Use the vocabulary terms listed below to complete each statement in exercises 1–2.

cross products **percent proportion**

1. Solve a proportion using ______________________.

2. The equation $\frac{\text{part}}{\text{whole}} = \frac{\text{percent}}{100}$ is called the ______________________.

Objective 1 Use the percent proportion to find the part.

Use the percent proportion to find the part. Round to the nearest tenth, if necessary.

1. 20% of 1400 **1.** ________________

2. 9% of 42 **2.** ________________

Use multiplication to find the part. Round to the nearest tenth, if necessary.

3. 175% of 50 **3.** ________________

4. 39.4% of 300 **4.** ________________

5. 0.7% of 3500 **5.** ________________

Solve each application problem. Round to the nearest tenth, if necessary.

6. A library has 330 visitors on Saturday, 20% of whom are children. How many are children? **6.** ________________

Name: Date:
Instructor: Section:

7. Bonnie Rae spent 15% of her savings on textbooks. If her savings were $560, find the amount that she spent on textbooks.

7. ______________

8. A survey at an intersection found that of 2200 drivers, 43% were wearing seat belts. How many drivers in the survey were wearing seat belts?

8. ______________

9. A family of four with a monthly income of $2100 spends 90% of its earnings and saves the balance. How much does the family save in one month?

9. ______________

10. In the last election, 74% of the eligible people actually voted. If there were 7844 voters, how many people were eligible?

10. ______________

Objective 2 Find the whole using the percent proportion.

Use the percent proportion to find the whole. Round to the nearest tenth, if necessary.

11. 15 is 5% of what number?

11. ______________

12. 36% of what number is 75?

12. ______________

13. 550 is 110% of what number?

13. ______________

14. 4.6% of what number is 69?

14. ______________

Name: Date:
Instructor: Section:

15. 24.5 is 0.7% of what number? **15.** ______________

Solve each application problem. Round to the nearest tenth, if necessary.

16. Michael Elders owns stock worth $4250, which is 17% of the value of his investments. What is the value of his investments? **16.** ______________

17. This year, there are 960 scholarship applications, which is 120% of the number of applications last year. Find the number of applications last year. **17.** ______________

18. Kathy Wicklund's overtime pay is $420, which is 12% of her total pay. What is her total pay? **18.** ______________

19. In one chemistry class, 60% of the students passed. If 90 students passed, how many students were in the class? **19.** ______________

20. On campus this semester there are 2028 married students, which is 26% of the total enrollment. Find the total enrollment. **20.** ______________

Objective 3 Find the percent using the percent proportion.

Use the percent proportion to find the whole. Round to the nearest tenth, if necessary.

21. What percent of 8000 is 4? **21.** ______________

Name: Date:
Instructor: Section:

22. 7 is what percent of 280? **22.** ________________

23. 650 is what percent of 13? **23.** ________________

24. What percent of 4.5 is 3.9? **24.** ________________

25. 550 is what percent of 1000? **25.** ________________

Solve each application problem. Round to the nearest tenth, if necessary.

26. In one shipment, 695 out of 27,800 crates were damaged. What percent of the crates were damaged? **26.** ________________

27. G&G Pharmacy has a total payroll of $89,350, of which $19,657 goes towards employee fringe benefits. What percent of the total payroll goes to fringe benefits? **27.** ________________

28. Vera's Antique Shoppe says that of its 5100 items in stock, 4233 are just plain junk, while the rest are antiques. What percent of the number of items in stock is antiques? **28.** ________________

29. In a motor cross, the leader has completed 108.8 miles of the 128-mile course. What percent of the total course has she completed? **29.** ________________

30. This month's class goal for Easy Writer Pen Company is 1,844,500 ballpoint pens. If 239,785 pens have been sold, what percent of the goal has been reached? **30.** ________________

Name: Date:
Instructor: Section:

Chapter 6 PERCENT

6.5 Using the Percent Equation

Learning Objectives
1 Use the percent equation to find the part.
2 Find the whole using the percent equation.
3 Find the percent using the percent equation.

Key Terms

Use the vocabulary terms listed below to complete each statement in exercises 1–2.

percent equation **percent**

1. A number written with a ______________ sign means "divided by 100".
2. The ____________________ is part = percent · whole.

Objective 1 Use the percent equation to find the part.

Find the part using the percent equation. Round to the nearest tenth, if necessary.

1. 70% of 920 1. ______________
2. 9% of 240 2. ______________
3. 140% of 76 3. ______________
4. 12.4% of 8100 4. ______________
5. 0.4% of 350 5. ______________
6. 125% of 76 6. ______________

Solve each application problem. Round to the nearest tenth or cent, if necessary.

7. A gardener has 56 clients, 25% of whom are residential. Find the number that are residential. 7. ______________

Name: Date:
Instructor: Section:

8. The total in sales at Hill's Market last month was \$87,428. If the profit was $1\frac{1}{2}$% of the sales, how much was the profit?

8. ______________

9. The sales tax rate in New York City is 8.375%. How much is the sales tax on a computer that costs \$1200?

9. ______________

10. A pair of shoes is marked 20% off. If the original price was \$56, how much is the discount?

10. ______________

Objective 2 Find the whole using the percent equation.

Find the whole using the percent equation. Round to the nearest tenth, if necessary.

11. 64 is 40% of what number?

11. ______________

12. 75% of what number is 1125?

12. ______________

13. $12\frac{1}{2}$% of what number is 270?

13. ______________

14. 75 is $6\frac{1}{4}$% of what number?

14. ______________

15. 35 is 153% of what number?

15. ______________

Name: Date:
Instructor: Section:

16. 170% of what number is 1462? **16.** ________________

Solve each application problem. Round to the nearest tenth or cent, if necessary.

17. A tank of an industrial chemical is 25% full. The tank now contains 160 gallons. How many gallons will it contain when it is full? **17.** ________________

18. Greg has completed 37.5% of the units needed for a degree. If he has completed 45 units, how many are needed for a degree? **18.** ________________

19. One day last week, 5% of the employees in a company were absent. If 25 employees were absent, how many employees are there? **19.** ________________

20. Over three months, the NASDAQ composite stock index dropped about 13.25%, or 330 points. What was its value at the beginning of the three month period? (Round your answer to the nearest dollar.) **20.** ________________

Objective 3 Find the percent using the percent equation.

Find the percent using the percent equation. Round to the nearest tenth, if necessary.

21. 15 is what percent of 75? **21.** ________________

22. What percent of 250 is 112.5? **22.** ________________

23. What percent of 160 is 8? **23.** ________________

Name: Date:
Instructor: Section:

24. What percent of 90 is 1.35? **24.** ______________

25. What percent of 18 is 44? **25.** ______________

26. What percent of 27 is 90? **26.** ______________

Solve each application problem. Round to the nearest tenth or cent, if necessary.

27. The Robinson family earns $2800 per month and saves $700 per month. What percent of the income is saved? **27.** ______________

28. The Hogan family drove 145 miles of their 500-mile vacation. What percent of the total number of miles did they drive? **28.** ______________

29. Jane eats 1500 calories a day. If she eats 350 calories for breakfast, what percent of her daily calories is her breakfast? **29.** ______________

30. A house costs $225,000. The Lee's paid $45,000 as a down payment. What percent of the cost of the house is their down payment? **30.** ______________

Name: Date:
Instructor: Section:

Chapter 6 PERCENT

6.6 Solving Application Problems with Percent

Learning Objectives
1. Find sales tax.
2. Find commissions.
3. Find the discount and sale price.
4. Find the percent of change.

Key Terms

Use the vocabulary terms listed below to complete each statement in exercises 1–4.

sales tax **commission** **discount**

percent of increase or decrease

1. ________________ is a percent of the dollar value of total sales paid to a salesperson.

2. In a __________________________ problem, the increase or decrease is a percent of the original amount.

3. The percent of the total sales charged as tax is called the ________________.

4. The percent of the original price that is deducted from the original price is called the ______________________.

Objective 1 Find sales tax.

Find the amount of sales tax and the total cost. Round answers to the nearest cent, if necessary.

	Amount of sale	Tax Rate	
1.	$50	7%	1. Tax ____________ Total ____________
2.	$350	6.5%	2. Tax ____________ Total ____________
3.	$67	9%	3. Tax ____________ Total ____________

Name: Date:
Instructor: Section:

Find the sales tax rate. Round answers to the hundredth, if necessary.

	Amount of sale	**Amount of Tax**	
4.	$450	$36	**4.** ______________
5.	$215	$10.75	**5.** ______________
6.	$78	$1.17	**6.** ______________

Solve each application problem. Round money answers to the nearest cent, if necessary.

7. A television set sells for $750 plus 8% sales tax. Find the price of the set including sales tax. **7.** ______________

8. A gold bracelet costs $1300 not including a sales tax of $71.50. Find the sales tax rate. **8.** ______________

Objective 2 Find commissions.

Find the commission earned. Round answers to the nearest cent, if necessary.

	Amount of sale	**Rate of Commission**	
9.	$6225	2.5%	**9.** ______________
10.	$156,000	3%	**10.** ______________

Name: Date:
Instructor: Section:

11. $75,000 4% **11.** ______________

Find the rate of commission. Round answers to the hundredth, if necessary.

	Amount of Sale	Amount of Commission	
12.	$3200	$480	**12.** ______________
13.	$5783	$231.32	**13.** ______________
14.	$25,000	$3750	**14.** ______________

Solve each application problem. Round money answers to the nearest cent, if necessary.

15. Nicole had sales of $18,306 in the month of October. If her rate of commission is 12%, find the amount of commission that she earned. **15.** ______________

16. A business property has just been sold for $1,692,804. The real estate agent selling the property earned a commission of $42,320.10. Find the rate of commission. **16.** ______________

Name: Date:
Instructor: Section:

Objective 3 Find the discount and sale price.

Find the amount of discount and the amount paid after the discount. Round money answers to the nearest cent, if necessary.

	Original price	Rate of Discount	
17.	\$200	15%	**17. Discount** ________ **Amount paid** ________
18.	\$595.80	20%	**18. Discount** ________ **Amount paid** ________
19.	\$205.50	5%	**19. Discount** ________ **Amount paid** ________
20.	\$24.95	60%	**20. Discount** ________ **Amount paid** ________

Solve each application problem. Round money answers to the nearest cent, if necessary.

21. Mike Lee can purchase a new car at 8% below window sticker price. Find the amount he can save on a car with a window sticker price of \$17,608.

21. ________________

22. A "Super 35% Off Sale" begins today. What is the price of a hair dryer normally priced at \$15?

22. ________________

Name: Date:
Instructor: Section:

23. Geishe's Shoes sells shoes at 33% off the regular price. Find the price of a pair of shoes normally priced at $54, after the discount is given.

23. ______________

Objective 4 Find the percent of change.

Solve each application problem. Round to the nearest tenth of a percent, if necessary.

24. Enrollment in secondary education courses increased from 1900 students last semester to 2280 students this semester. Find the percent of increase.

24. ______________

25. The number of days employees of Prodex Manufacturing Company were absent from their jobs decreased from 96 days last month to 72 days this month. Find the percent of decrease.

25. ______________

26. The earnings per share of Amy's Cosmetic Company decreased from $1.20 to $0.86 in the last year. Find the percent of decrease.

26. ______________

27. The membership of Pleasant Acres Golf Club was 320 two years ago. The membership is now 740. Find the percent of increase in the two years.

27. ______________

Name: Date:
Instructor: Section:

28. The price of a certain model of calculator was \$33.50 five years ago. This calculator now costs \$18.75. Find the percent of decrease in the price in the last five years.

28. ________________

29. In 1980, there were approximately 3,612,000 births in the U.S. In 2002, there were approximately 4,022,000 births in the U.S. Find the percent of increase.

29. ________________

30. One day in 2008, the Dow Jones Industrial Average dropped from about 12,635 to 12,265. Find the percent of decrease.

30. ________________

Name: Date:
Instructor: Section:

Chapter 6 PERCENT

6.7 Simple Interest

Learning Objectives
1 Find the simple interest on a loan.
2 Find the total amount due on a loan.

Key Terms

Use the vocabulary terms listed below to complete each statement in exercises 1–5.

interest **interest formula** **simple interest** **principal**

rate of interest

1. The charge for money borrowed or loaned, expressed as a percent, is called ________________.

2. A fee paid for borrowing or lending money is called ____________________.

3. The formula $I = p \cdot r \cdot t$ is called the ____________________________.

4. Use the formula $I = p \cdot r \cdot t$ to compute the amount of ______________________ due on a loan.

5. The amount of money borrowed or loaned is called the ____________________.

Objective 1 Find the simple interest on a loan.

Find the interest. Round to the nearest cent, if necessary.

	Principal	Rate	Time in Years	
1.	\$400	2%	3	1. ________________
2.	\$80	5%	1	2. ________________
3.	\$5280	8%	5	3. ________________
4.	\$780	10%	$2\frac{1}{2}$	4. ________________

Name: Date:
Instructor: Section:

	Principal	**Rate**	**Time in Years**	
5.	\$360	6%	$1\frac{1}{2}$ years	**5.** ____________
6.	\$620	16%	$1\frac{1}{4}$ years	**6.** ____________

Find the interest. Round to the nearest cent, if necessary.

	Principal	**Rate**	**Time in Months**	
7.	\$200	16%	3	**7.** ____________
8.	\$500	11%	9	**8.** ____________
9.	\$820	3%	18	**9.** ____________
10.	\$522	8%	21	**10.** ____________
11.	\$2000	12%	39	**11.** ____________
12.	\$14,400	7%	7	**12.** ____________

Solve each application problem. Round to the nearest cent, if necessary.

13. Diane lends \$6500 for 18 months at 12%. How much interest will she earn? **13.** ____________

14. A mother lends \$6500 to her daughter for 15 months and charges 9% interest. Find the interest charged on the loan. **14.** ____________

Name: Date:
Instructor: Section:

15. Kareem invests $1500 at 16% for 6 months. What amount of interest will he earn?

15. ______________

16. Darla deposits $680 at 14% for 1 year. How much interest will she earn?

16. ______________

17. A savings account pays $2\frac{1}{2}\%$ interest per year. How much interest will be earned on $850 invested for 3 years?

17. ______________

Objective 2 Find the total amount due on a loan.

Find the total amount due on each loan. Round to the nearest cent, if necessary.

	Principal	Rate	Time	
18.	$200	11%	1 year	**18.** ______________
19.	$3000	5%	6 months	**19.** ______________
20.	$1500	8%	18 months	**20.** ______________
21.	$900	10%	$2\frac{1}{2}$ years	**21.** ______________
22.	$6000	7%	5 months	**22.** ______________

Name: Date:
Instructor: Section:

	Principal	Rate	Time	
23.	\$15,400	16%	5 years	**23.** ______________
24.	\$18,200	7%	8 months	**24.** ______________
25.	\$30,900	16%	4 months	**25.** ______________

Solve each application problem. Round to the nearest cent, if necessary.

26. A loan of \$1500 will be paid back with 12% interest at the end of 27 months. Find the total amount due.

26. ______________

27. An employee credit union pays 7% interest. If Mario deposits \$2100 in his account for $\frac{1}{3}$ year and makes no withdrawals or further deposits, find the total amount in Mario's account after that time.

27. ______________

28. An investor deposits \$7000 at 16% for 2 years. If there are no withdrawals or further deposits, find the total amount in the account after 2 years.

28. ______________

29. Mary Ann borrows \$1200 at 10% for 3 months. Find the total amount due.

29. ______________

30. Cheri borrowed \$45,000 for 9 months at 11.2% interest. Find the total amount she must repay.

30. ______________

Name: Date:
Instructor: Section:

Chapter 6 PERCENT

6.8 Compound Interest

Learning Objectives

1	Understand compound interest.
2	Understand compound amount.
3	Find the compound amount.
4	Use a compound interest table.
5	Find the compound amount and the amount of compound interest.

Key Terms

Use the vocabulary terms listed below to complete each statement in exercises 1–3.

compound interest **compound amount** **compounding**

1. Interest paid on principal plus past interest is called ______________.

2. The total amount in an account, including compound interest and the original principal, is called the ____________________.

3. When the amount of interest is computed based on the principal plus the past interest, use a process called ______________________.

Objective 1 Understand compound interest.
Objective 2 Understand compound amount.

1. Belinda deposited $2000 in an account earning 5% annually. How much is in the account at the end of the first year?

1. ________________

2. If Belinda makes no withdrawals, how much money is in her account at the end of two years?

2. ________________

3. If Belinda makes no withdrawals, how much money is in her account at the end of three years? How much interest has she earned in total?

3. ________________

____

Name: Date:
Instructor: Section:

Objective 3 Find the compound amount.

Find the compound amount given the following deposits. Interest is compounded annually. Round to the nearest cent, if necessary.

4. $7000 at 5% for 3 years **4.** ________

5. $4500 at 4% for 2 years **5.** ________

6. $3200 at 7% for 2 years **6.** ________

7. $24,600 at 5% for 4 years **7.** ________

8. $8000 at 8% for 4 years **8.** ________

9. $3200 at 6% for 2 years **9.** ________

10. $1200 at 2% for 3 years **10.** ________

Name: Date:
Instructor: Section:

Objective 4 Use a compound interest table.

Use the table for compound interest to find the compound amount. Interest is compounded annually. Round to the nearest cent, if necessary.

Time Periods	3.00%	3.50%	4.00%	4.50%	5.00%	5.50%	6.00%	8.00%	Time Periods
1	1.0300	1.0350	1.0400	1.0450	1.0500	1.0550	1.0600	1.0800	1
2	1.0609	1.0712	1.0816	1.0920	1.1025	1.1130	1.1236	1.1664	2
3	1.0927	1.1087	1.1249	1.1412	1.1576	1.1742	1.1910	1.2597	3
4	1.1255	1.1475	1.1699	1.1925	1.2155	1.2388	1.2625	1.3605	4
5	1.1593	1.1877	1.2167	1.2462	1.2763	1.3070	1.3382	1.4693	5
6	1.1941	1.2293	1.2653	1.3023	1.3401	1.3788	1.4185	1.5869	6
7	1.2299	1.2723	1.3159	1.3609	1.4071	1.4547	1.5036	1.7138	7
8	1.2668	1.3168	1.3686	1.4221	1.4775	1.5347	1.5938	1.8509	8
9	1.3048	1.3629	1.4233	1.4861	1.5513	1.6191	1.6895	1.9990	9
10	1.3439	1.4106	1.4802	1.5530	1.6289	1.7081	1.7908	2.1589	10
11	1.3842	1.4600	1.5395	1.6229	1.7103	1.8021	1.8983	2.3316	11
12	1.4258	1.5111	1.6010	1.6959	1.7959	1.9012	2.0122	2.5182	12

11. $1000 at 6% for 4 years **11.** ________________

12. $4000 at 5% for 9 years **12.** ________________

13. $7500 at 6% for 7 years **13.** ________________

14. $60 at 5.5% for 2 years **14.** ________________

15. $48 at 8% for 3 years **15.** ________________

Name: Date:
Instructor: Section:

16. $8428.17 at $4\frac{1}{2}$% for 6 years **16.** ______________

17. $10,422.75 at $5\frac{1}{2}$% for 12 years **17.** ______________

18. $24,600 at 5% for 4 years **18.** ______________

Objective 5 Find the compound amount and the amount of compound interest.

Find the compound amount and the compound interest. Round to the nearest cent, if necessary. Use the table for compound interest in your text book to find the compound amount. Interest is compounded annually.

	Principal	Rate	Time in Years	
19.	$1000	$3\frac{1}{2}$%	7	**19.** **Amount** ______________ **Interest** ______________
20.	$8500	6%	12	**20.** **Amount** ______________ **Interest** ______________
21.	$12,800	$5\frac{1}{2}$%	9	**21.** **Amount** ______________ **Interest** ______________
22.	$9150	8%	8	**22.** **Amount** ______________ **Interest** ______________
23.	$21,400	$4\frac{1}{2}$%	11	**23.** **Amount** ______________ **Interest** ______________

Name: Date:
Instructor: Section:

	Principal	Rate	Time in Years	
24.	$45,000	4%	4	**24.** **Amount** ____________ **Interest**____________
25.	$78,000	3%	12	**25.** **Amount** ____________ **Interest**____________
26.	$8000	3%	12	**26.** **Amount** ____________ **Interest**____________
27.	$1000	4.5%	6	**27.** **Amount** ____________ **Interest**____________
28.	$8500	8%	12	**28.** **Amount** ____________ **Interest**____________

Solve each application problem. Use the table for compound interest in your text book to find the compound amount. Interest is compounded annually. Round to the nearest cent, if necessary.

29. Scott lends $9000 to the owner of a new restaurant. He will be repaid at the end of 6 years at 8% interest compounded annually. Find how much he will be repaid and how much interest he will earn.

29.
Amount ____________

Interest____________

30. Michelle invests $2500 in a health spa. She will be repaid at the end of 5 years at 6% interest compounded annually. Find how much she will be repaid and how much interest she will earn.

30.
Amount ____________

Interest____________

Name: Date:
Instructor: Section:

Chapter 7 MEASUREMENT

7.1 Problem Solving with U.S. Customary Measurements

Learning Objectives

1. Learn the basic U.S. customary measurement units.
2. Convert among measurement units using multiplication or division.
3. Convert among measurement units using unit fractions.
4. Solve application problems using U.S. customary measurement units.

Key Terms

Use the vocabulary terms listed below to complete each statement in exercises 1–3.

U.S. customary measurement units **unit fractions** **metric system**

1. The ________________________ is based on units of ten.

2. ____________________________ are used to convert among different measurements.

3. _________________________________ include inches, feet, quarts, and pounds.

Objective 1 Learn the basic U.S. customary measurement units.

Fill in the blanks.

1. 1 ft = _____ in. **1.** ______________

2. _____ oz = 1 lb **2.** ______________

3. _____ pt = 1 qt **3.** ______________

4. 1 T = _____ lb **4.** ______________

5. 1 mi = _____ ft **5.** ______________

6. _____ qt = 1 gal **6.** ______________

7. 1 yd = _____ ft **7.** ______________

Name: Date:
Instructor: Section:

8. 1 c = _____ fl oz **8.** ______________

Objective 2 Convert among measurement units using multiplication or division.

Convert each measurement using multiplication or division.

9. 12 ft to yards **9.** ______________

10. 40 pt to gallons **10.** ______________

11. $17\frac{1}{2}$ ft to inches **11.** ______________

12. $3\frac{1}{2}$ lb to ounces **12.** ______________

13. 5 T to pound **13.** ______________

14. 27,720 ft to miles **14.** ______________

15. 75 sec to minutes **15.** ______________

16. 380 min to hours **16.** ______________

17. 30 in to yards **17.** ______________

Objective 3 Convert among measurement units using unit fractions.

Convert each measurement using unit fractions.

18. 12 yards to inches **18.** ______________

Name: Date:
Instructor: Section:

19. 28 pt to gallons **19.** ______________

20. 38 c to pints **20.** ______________

21. 60 oz to pounds **21.** ______________

22. 3000 lb to tons **22.** ______________

23. 4 mi to feet **23.** ______________

24. 12 cups to gallons **24.** ______________

25. 5 days to hours **25.** ______________

26. $3\frac{1}{4}$ gal to quarts **26.** ______________

Objective 4 Solve application problems using U.S. customary measurement units.

Solve each application problem using the six problem-solving steps.

27. Lee paid $4.65 for 14 oz of honey baked ham. What is the price per pound, to the nearest cent? **27.** ______________

Name: Date:
Instructor: Section:

28. Sweet Suzie's Shop makes 49,000 lb of fudge each week.

(a) How many tons of fudge are produced each day of the 7-day week?

(b) How many tons of fudge are produced each year?

28. (a) ____________

(b) ____________

29. Clarissa paid \$1.79 for 4.5 oz of nuts. What is the cost per pound, to the nearest cent?

29. ________________

30. At a preschool, each of 20 children drinks about $\frac{3}{4}$ c of juice with their snack each day. The school is open 3 days a week. How many quarts of juice are needed for 2 weeks of snacks?

30. ________________

Name: Date:
Instructor: Section:

Chapter 7 MEASUREMENT

7.2 The Metric System—Length

Learning Objectives
1 Learn the basic metric units of length.
2 Use unit fractions to convert among units.
3 Move the decimal point to convert among units.

Key Terms

Use the vocabulary terms listed below to complete each statement in exercises 1–3.

meter **prefix** **metric conversion line**

1. Attaching a ______________________ such as "kilo-" or "milli" to the words "meter", "liter", or "gram" gives the names of larger or smaller units.

2. A line showing the various metric measurement prefixes and their size relationship to each other is called a ____________________________________.

3. The basic unit of length in the metric system is the ____________________.

Objective 1 Learn the basic metric units of length.

Choose the most reasonable metric unit. Choose from **km**, **m**, **cm**, *or* **mm**.

1. the thickness of a dime **1.** ________________

2. the width of an adult's finger **2.** ________________

3. the width of a twin bed **3.** ________________

4. the distance between two cities **4.** ________________

5. the length of a room **5.** ________________

6. the width of a textbook **6.** ________________

7. the thickness of a piece of cardboard **7.** ________________

8. the length of a nail **8.** ________________

9. the distance driven in 2 hours **9.** ________________

10. the height of a giant redwood tree **10.** ________________

Name: Date:
Instructor: Section:

Objective 2 Use unit fractions to convert among units.

Convert each measurement using unit fractions.

11. 7 cm to meters **11.** ______________

12. 25.87 m to centimeters **12.** ______________

13. 2.3 m to millimeters **13.** ______________

14. 53.1 m to centimeters **14.** ______________

15. 450 m to kilometers **15.** ______________

16. 140 millimeters to meters **16.** ______________

Solve each application problem.

17. A dog is 602 mm in length. Give its length in centimeters. **17.** ______________

18. Leon's waist size is 72 cm. Give his waist size in millimeters. **18.** ______________

19. Is 103 cm more or less than 1 m? What is the difference in lengths? **19.** ______________

Name: Date:
Instructor: Section:

20. Is 23 km more or less than 2311 m? What is the difference in lengths?

20. ______________

Objective 3 Move the decimal point to convert among units.

Convert each measurement using the metric conversion line.

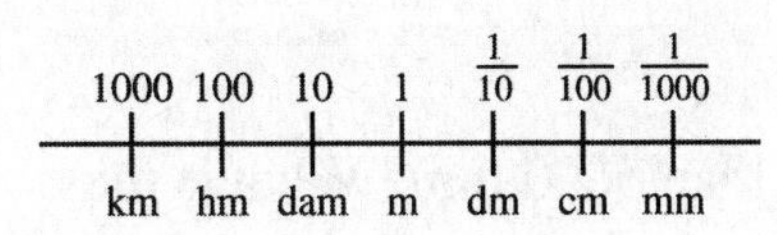

21. 63.6 cm to meters

21. ______________

22. 1.94 cm to millimeters

22. ______________

23. 14,500 m to kilometers

23. ______________

24. 10.35 km to meters

24. ______________

25. 3.5 cm to kilometers

25. ______________

26. 2.86 km to centimeters

26. ______________

Solve each application problem.

27. A driver is told to turn left in 0.61 km. How many meters is this?

27. ______________

Name: Date:
Instructor: Section:

28. A building is 83.6 m tall. How many kilometers is this? **28.** _______________

29. Is 4.72 m more or less than 271 cm? What is the difference in lengths? **29.** _______________

30. Is 5.38 m more or less than 5000 mm? What is the difference in lengths? **30.** _______________

Name: Date:
Instructor: Section:

Chapter 7 MEASUREMENT

7.3 The Metric System—Capacity and Weight (Mass)

Learning Objectives
1 Learn the basic metric units of capacity.
2 Convert among metric capacity units.
3 Learn the basic metric units of weight (mass).
4 Convert among metric weight (mass) units.
5 Distinguish among basic metric units of length, capacity, and weight (mass).

Key Terms

Use the vocabulary terms listed below to complete each statement in exercises 1–2.

liter **gram**

1. The basic unit of weight (mass) in the metric system is the ________________.

2. The basic unit of capacity in the metric system is the ____________________.

Objective 1 Learn the basic metric units of capacity.

Choose the most reasonable metric unit. Choose from **L**, *or* **ml**.

1. the amount of soda in a can 1. ________________

2. the amount of water in a bathtub 2. ________________

3. the amount of orange juice in a large bottle 3. ________________

4. the amount of cough syrup in one dose 4. ________________

Objective 2 Convert among metric capacity units.

Convert each measurement. Use unit fractions or the metric conversion line.

5. 7 L to kiloliters 5. ________________

6. 9.7 L to milliliters 6. ________________

7. 2.5 L to milliliters 7. ________________

Name: Date:
Instructor: Section:

8. 32.4 kL to milliliters **8.** ________

9. 836 kL to liters **9.** ________

10. 523 mL to liters **10.** ________

11. 7863 mL to liters **11.** ________

12. 7724 mL to kiloliters **12.** ________

Objective 3 Learn the basic metric units of weight (mass).

Choose the most reasonable metric unit. Choose from **kg, g,** *or* **mg.**

13. the weight (mass) of a vitamin pill **13.** ________

14. the weight (mass) of a car **14.** ________

15. the weight (mass) of an apple **15.** ________

16. the weight (mass) of an egg **16.** ________

Objective 4 Convert among metric weight (mass) units.

Convert each measurement. Use unit fractions or the metric conversion line.

17. 9000 g to kilograms **17.** ________

18. 27,000 g to kilograms **18.** ________

Name: Date:
Instructor: Section:

19. 6.3 kg to grams **19.** ______________

20. 0.76 kg to grams **20.** ______________

21. 4.7 g to milligrams **21.** ______________

22. 4.91 kg to milligrams **22.** ______________

23. 8745 mg to kilograms **23.** ______________

24. 42 mg to grams **24.** ______________

Objective 5 Distinguish among basic metric units of length, capacity, and weight (mass).

Choose the most reasonable metric unit. Choose from ***km, m, cm, mm, mL, L, kg, g,*** *or* ***mg***.

25. The tablet contains 200 _____ of aspirin. **25.** ______________

26. Buy a 5 _____ bottle of water. **26.** ______________

27. She drove 450 _____ in one day. **27.** ______________

28. The piece of wood is 20 _____ wide. **28.** ______________

29. The piece of wood weighs 5 __________. **29.** ______________

30. A paperclip is 3 _________ long. **30.** ______________

Name: Date:
Instructor: Section:

Chapter 7 MEASUREMENT

7.4 Problem Solving with Metric Measurement

Learning Objectives
1 Solve application problems involving metric measurements.

Key Terms

Use the vocabulary terms listed below to complete each statement in exercises 1–3.

meter **liter** **gram**

1. A ____________________ is the weight of 1 mL of water.
2. A ____________________ is a little longer than a yard.
3. A ____________________ is a little more than one quart.

Objective 1 Solve application problems involving metric measurements.

Solve each application problem. Round money answers to the nearest cent.

1. How many 150 mL servings are in 9 L of juice? **1.** ______________

2. A commodity costs $0.85 per kilogram. Find the cost of 3 kg 70 g. Round your answer to the nearest cent. **2.** ______________

3. Metal chain costs $5.26 per meter. Find the cost of 2 m 47 cm of the chain. Round your answer to the nearest cent. **3.** ______________

4. In a laboratory each mouse requires 25 g of food per day. How many kilograms of food are needed to feed 357 mice each day? **4.** ______________

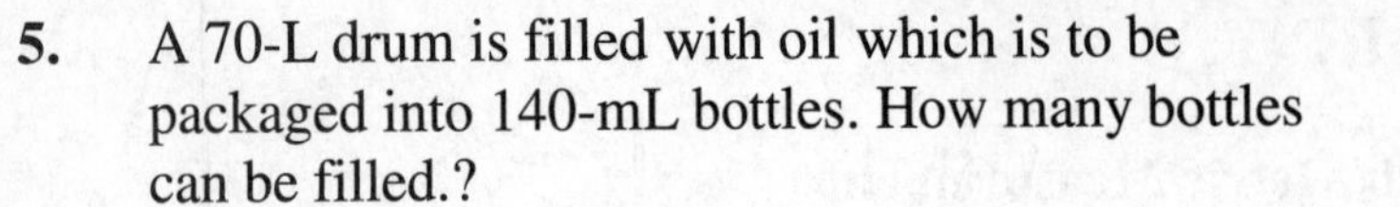
Name: Date:
Instructor: Section:

5. A 70-L drum is filled with oil which is to be packaged into 140-mL bottles. How many bottles can be filled.?

5. ________________

6. A patio garden slab measures 50 cm by 50 cm by 5 cm and weighs 80 kg. How many kilograms would a truck load of 70 slabs weight?

6. ________________

7. Helene has 3 m 47 cm of red fabric left from one project and 4 m 86 cm of the same fabric from another project. Find the total amount of red fabric she has left in meters.

7. ________________

8. A boy weighed 4 kg 82 g at birth. A month later he weighted 6 kg 17 g. How much weight had he gained, in kilograms?

8. ________________

9. If you drink 175 mL of soda pop every day for two weeks, how many liters would you consume in this time period?

9. ________________

10. If 1.8 kg of candy is to be divided equally among 9 children, how many grams will each child receive?

10. ________________

Name: Date:
Instructor: Section:

11. Vitamin C comes in pills with a strength of 500 mg. How many pills do you need to take if you want a dosage of 1.5 g?

11. ________________

12. A ball of yard weighs 140 g. Ellie knitted a sweater for her boyfriend that used 11 balls of yarn. How many kilograms did the finished sweater weigh?

12. ________________

13. A fish tank can hold up to 75.6 L of water. If there are 70,000 mL of water in the tank, how many more milliliters of water can the tank hold?

13. ________________

14. A dollar bill weighs approximately 1g.

(a) How many kg will $1,000,000 in $1 bills weigh?

(b) How many kg will $1,000,000 in $100 bills weigh?

14. (a) ______________

(b) ______________

15. Henry's bowls with a 7-kg bowling ball, while Denise uses a ball that weighs 5 kg 750g. How much heavier is Henry's bowling ball than Denise's?

15. ________________

16. Ari switched from drinking two 8-ounce cups of double espresso per day to three 8-ounce cups of drip coffee per day. If each 8-ounce cup of double

16. ________________

espresso contains 160 mg of caffeine and each 8-ounce cup of drip coffee contains 90 mg of caffeine, did Ari reduce or increase the amount of caffeine he drinks each day? How many milligrams were the decrease/increase?

17. A cookie recipe for 24 cookies uses 140 g sugar. How many grams of sugar re needed to make 144 cookies?

17. ______________

18. Lucy purchased 1 m 60 cm of fabric at $6.25 per meter for a jacket and 1m 80 cm of fabric at $4.75 per m for a dress. How much did she spend in total?

18. ______________

19. A certain drug is sold in a 5 L bottle. It is dispensed in 250 mL units. How many units are dispensed from the bottle?

19. ______________

20. The label on a bottle of pills says that there are 3.5 mg of the medication in 5 pills. If a patient needs to take 8.4 mg of the medication, how many pills does he need to take?

20. ______________

Name: Date:
Instructor: Section:

21. The speed limit on a road is 80 kilometers per hour. Lee drove 258 km in three hours. Was his average speed above the speed limit, at the speed limit, or below the speed limit? By how many kilometers per hour?

21. ______________

22. Bulk beans are on sale at $0.85 per kilogram. Rochelle scooped some beans into a bag and put it on the scale. How much will she pay for 3 kg 20 g of beans, to the nearest cent?

22. ______________

23. Mary Beth needs 2 m 70 cm of material to make a skirt. The price is $6.49 per meter plus a 6% sales tax. How much will she pay, to the nearest cent?

23. ______________

24. Colby cheese is on sale at $9.89 per kilogram. Nelson bought 750 g of the cheese. How much did he pay, to the nearest cent?

24. ______________

25. Which case of pasta sauce is the better buy: a $29.50 case that holds ten 1 L bottles or a $28 case that holds sixteen 600 mL bottles? What is the price per liter?

25. ______________

26. Jannella's doctor wants her to take 2.8 g of medication each day in four equal doses. How many milligrams should be in each dose?

26. ______________

Name: Date:
Instructor: Section:

27. How far can a car that gets 14.5 km per liter of gasoline go on 13 liters of gasoline?

27. ________________

28. Shaina gained 13 kg 636 g during her pregnancy. If she weighed 147 kg 400 g before she became pregnant, how much did she weigh at the end of her pregnancy?

28. ________________

29. John's car gets 14.5 km per liter of gasoline, and the gas tank can hold 62 L of gasoline. He wants to drive 1000 km. Will he need to stop to buy gas? If so, how many liters of gas will he need? Round answer to the nearest whole number.

29. ________________

30. A garden measures 362 cm by 554 cm. If chicken wire fencing costs $2.00 per meter, how much will it cost for enough fencing to completely surround the garden?

30. ________________

Name: Date:
Instructor: Section:

Chapter 7 MEASUREMENT

7.5 Metric–U.S. Customary Conversions and Temperature

Learning Objectives

1 Use unit fractions to convert between metric and U.S. customary units.
2 Learn common temperatures on the Celsius scale.
3 Use formulas to convert between Celsius and Fahrenheit temperatures.

Key Terms

Use the vocabulary terms listed below to complete each statement in exercises 1–2.

Celsius **Fahrenheit**

1. The ____________________ scale is used to measure temperature in the metric system.

2. The ____________________ scale is used to measure temperature in the U.S. customary system.

Objective 1 Use unit fractions to convert between metric and U.S. customary units.

Use the table in your textbook and unit fractions to make the following conversions. Round answers to the nearest tenth.

1. 9.68 kg to pounds **1.** ________________

2. 46.8 L to quarts **2.** ________________

3. 5.2 yd to meters **3.** ________________

4. 12.9 m to feet **4.** ________________

5. 22.5 pounds to kilograms **5.** ________________

6. 20 gallons to liters **6.** ________________

7. 16 in. to centimeters **7.** ________________

Name: Date:
Instructor: Section:

Solve each application problem. Round to the nearest tenth or cent, if necessary.

8. Suppose you decide to put together a do-it-yourself picture frame that measures 24 cm by 30 cm. The wood for the frame costs $1.40 per foot. Find the approximate cost of the wood. Round your answer to the nearest cent. (Hint: Start by finding the number of meters of wood in the frame.)

8. ______________

9. A recipe calls for 2.5 L of chicken broth. How many quarts of chicken broth should be used to make this recipe?

9. ______________

10. The distance from Pittsburgh to Philadelphia is 291 mi. Find the distance in kilometers.

10. ______________

11. If paint sells for $11 per gal, find the cost of four liters.

11. ______________

12. Cheryl is making a dress for each of her twin nieces. Each dress requires 72 inches of lace trimming. If the lace costs $1.50 per meter, how much will the lace cost to the nearest cent?

12. ______________

13. A 3-L bottle of beverage sells for $2.90. A gallon bottle of the same beverage sells for $3.60. Which is the better value?

13. ______________

Name: Date:
Instructor: Section:

Objective 2 Learn common temperatures on the Celsius scale.

Choose the most reasonable temperature for each situation.

14. hot water in a bathtub
27ºC 40ºC 100ºC

14. ______________

15. fall day
13ºC 50ºC 65ºC

15. ______________

16. ice cream
−10ºC 4ºC 30ºC

16. ______________

17. Hot cocoa
65°C 65°F

17. ______________

18. Normal body temperature
37°C 37°F

18. ______________

19. Oven temperature
300°C 300°F

19. ______________

20. Spring day
50°C 50°F

20. ______________

Objective 3 Use formulas to convert between Celsius and Fahrenheit temperatures.

Use the conversions formulas and the order of operations to convert Fahrenheit temperatures to Celsius and Celsius temperatures to Fahrenheit. Round your answers to the nearest degree, if necessary.

21. 62ºF

21. ______________

22. 125ºF

22. ______________

23. 80ºF

23. ______________

Name: Date:
Instructor: Section:

24. 10°C **24.** ________________

25. 30°C **25.** ________________

26. 150°C **26.** ________________

Solve each application problem. Round to the nearest degree, if necessary.

27. A kiln for firing pottery reaches a temperature of 450°C. What is the temperature in degrees Fahrenheit? **27.** ________________

28. The highest temperature during a recent year in Phoenix was 115°F. What is the temperature in degrees Celsius? **28.** ________________

29. A recipe for roast beef calls for an oven temperature of 400°F. What is the temperature in degrees Celsius? **29.** ________________

30. The average July and August temperature in Baghdad, Iraq is 40°C –45°C. What is the temperature in degrees Fahrenheit? **30.** ________________

Name: Date:
Instructor: Section:

Chapter 8 GEOMETRY

8.1 Basic Geometric Terms

Learning Objectives

1. Identify and name lines, line segments, and rays.
2. Identify parallel and intersecting lines.
3. Identify and name angles.
4. Classify angles as right, acute, straight, or obtuse.
5. Identify perpendicular lines.

Key Terms

Use the vocabulary terms listed below to complete each statement in exercises 1–13.

point	**line**	**line segment**	**ray**
angle	**degrees**	**right angle**	**acute angle**
obtuse angle		**straight angle**	
intersecting lines	**perpendicular lines**		**parallel lines**

1. A ________________________ is a part of a line that has one endpoint and which extends infinitely in one direction.

2. Two lines that intersect to form a right angle are ________________________.

3. An angle whose measure is between 90º and 180º is an ____________________.

4. A ______________________ is a location in space.

5. Two rays with a common endpoint form an ________________________.

6. A set of points that form a straight path that extends infinitely in both directions is called a __________________________.

7. An angle that measures less than 90º is called an ______________________.

8. Angles are measured using ______________________.

9. An angle whose measure is exactly 180º is called a ______________________.

10. Two lines in the same plane that never intersect are ______________________.

11. A part of a line with two endpoints is a ______________________.

12. Two lines that cross at one point are ________________________.

13. An angle whose measure is exactly 90º is a __________________________.

Name: Date:
Instructor: Section:

Objective 1 Identify and name lines, line segments, and rays.

Identify each figure as a line, line segment, or ray, and name it.

1.

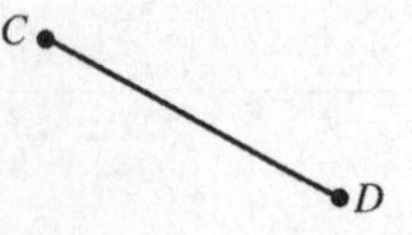

1. ______________

2.

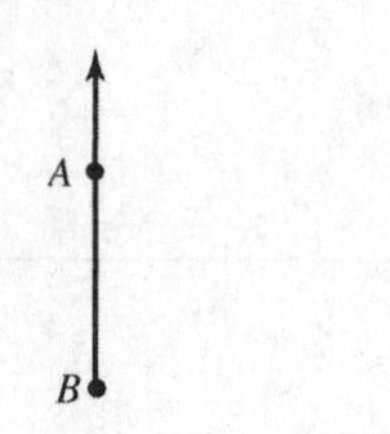

2. ______________

3. E F

3. ______________

4.

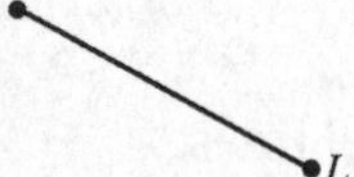

4. ______________

5.

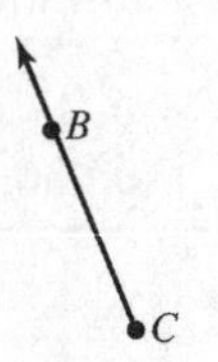

5. ______________

6.

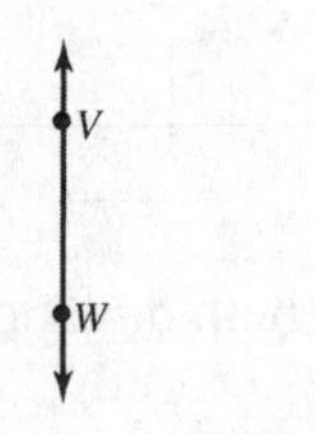

6. ______________

7. R S

7. ______________

Objective 2 Identify parallel and intersecting lines.

Label each pair of lines as appearing to be **parallel** *or* **intersecting.**

8.

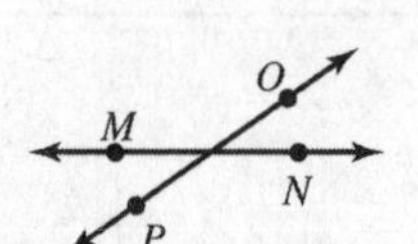

8. ______________

Name: Date:
Instructor: Section:

9.

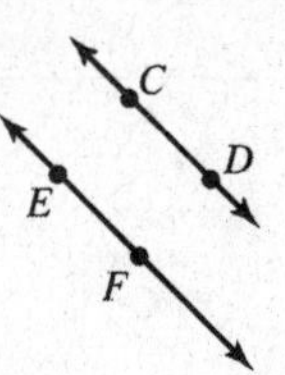

9. ______________

10.

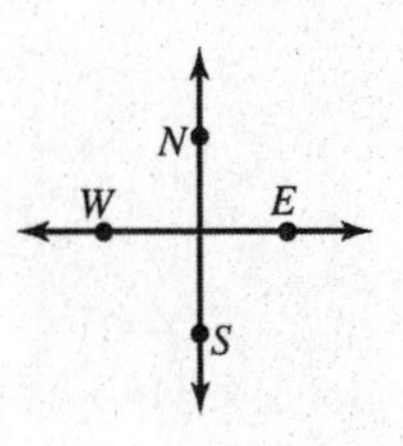

10. ______________

11.

11. ______________

Objective 3 Identify and name angles.

Name each angle drawn with darker rays by using the three-letter form of identification.

12.

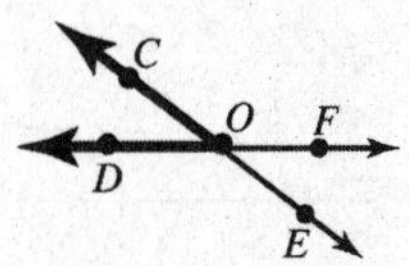

12. ______________

13.

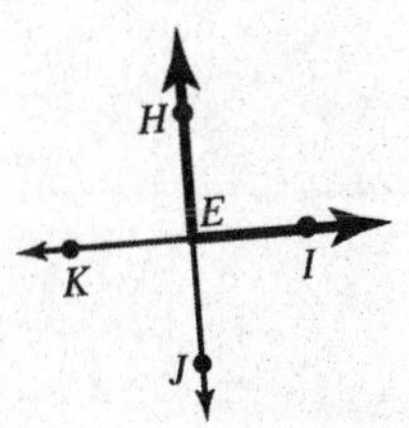

13. ______________

14.

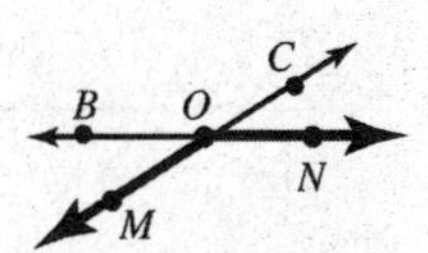

14. ______________

15.

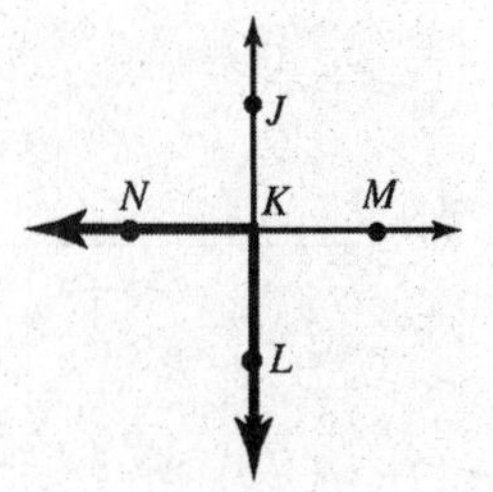

15. ______________

Name: Date:
Instructor: Section:

16. 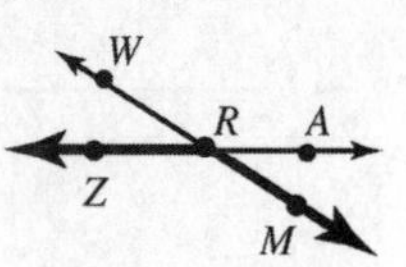

16. ____________

17. 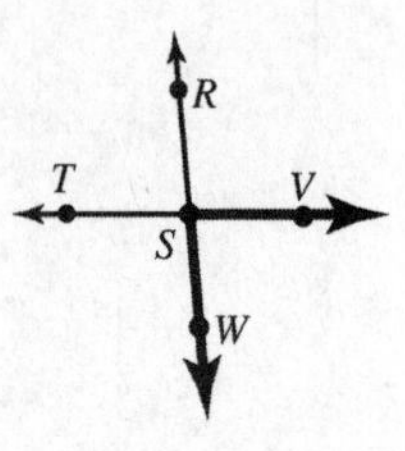

17. ____________

Objective 4 Classify angles as right, acute, straight, or obtuse.

Label each angle as **acute**, **right**, **obtuse**, *or* **straight**.

18. 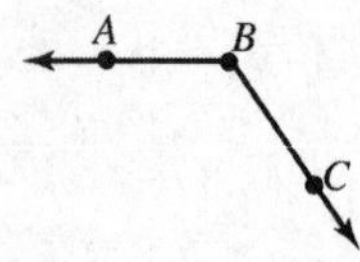

18. ____________

19. 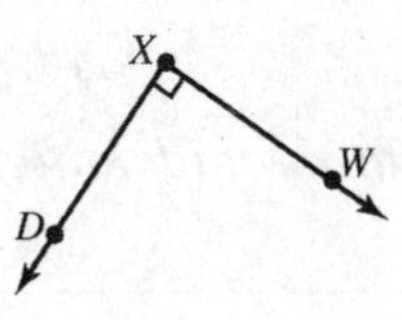

19. ____________

20. 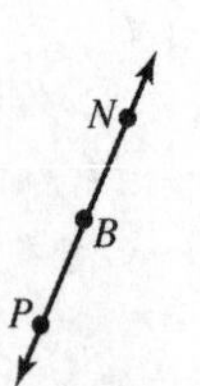

20. ____________

21. 

21. ____________

22. 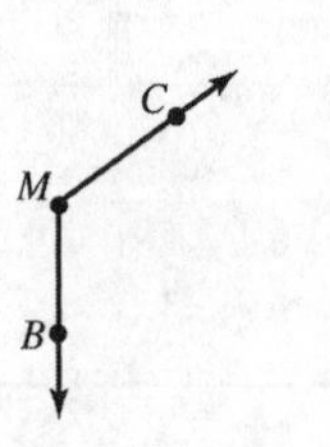

22. ____________

23. 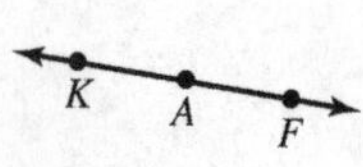

23. ____________

24. R S T

24. ____________

Name: Date:
Instructor: Section:

25.

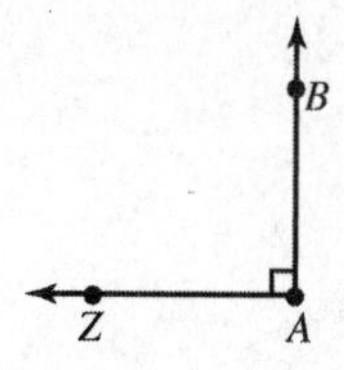

25. ______________

Objective 5 Identify perpendicular lines.

Label each pair of lines as appearing to be **parallel**, **perpendicular**, *or* **intersecting**.

26.

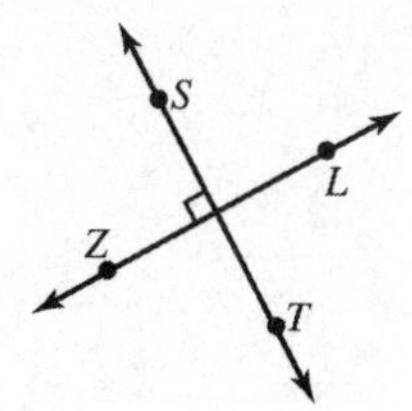

26. ______________

27.

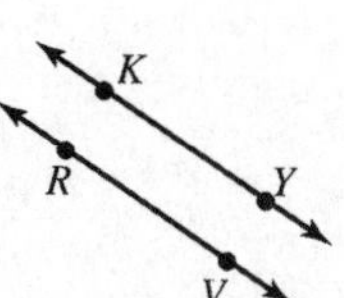

27. ______________

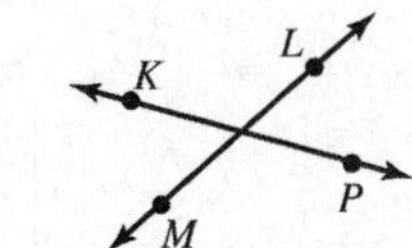

28. ______________

28.

29.

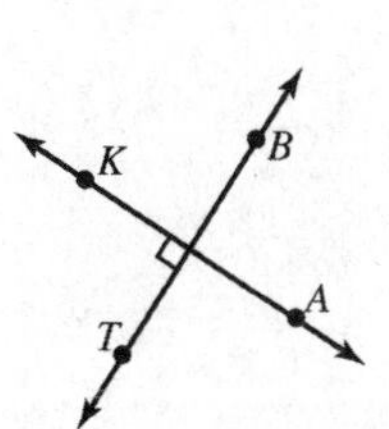

29. ______________

30.

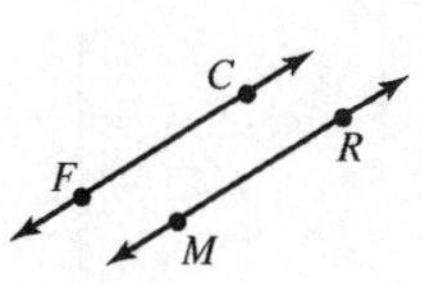

30. ______________

Name: Date:
Instructor: Section:

Chapter 8 GEOMETRY

8.2 Angles and Their Relationships

Learning Objectives

1. Identify complementary angles and supplementary angles and find the measure of complement or supplement of a given angle.
2. Identify congruent angles and vertical angles and use this knowledge to find the measures of angles.

Key Terms

Use the vocabulary terms listed below to complete each statement in exercises 1–4.

complementary angles **supplementary angles**

congruent angles **vertical angles**

1. The nonadjacent angles formed by two intersecting lines are called ______________________________.

2. Angles whose measures are equal are called ______________________________.

3. Two angles whose measures sum to 180° are ______________________________.

4. Two angles whose measures sum to 90° are ______________________________.

Objective 1 Identify complementary angles and supplementary angles and find the measure of complement or supplement of a given angle.

Find the complement of each angle.

1. 12° **1.** ________________

2. 43° **2.** ________________

3. 72° **3.** ________________

4. 66° **4.** ________________

5. 4° **5.** ________________

Find the supplement of each angle.

6. 121° **6.** ________________

7. 16° **7.** ________________

Name: Date:
Instructor: Section:

8. 168° **8.** ________

9. 38° **9.** ________

10. 90° **10.** ________

Identify each pair of complementary angles.

11. **11.** ________

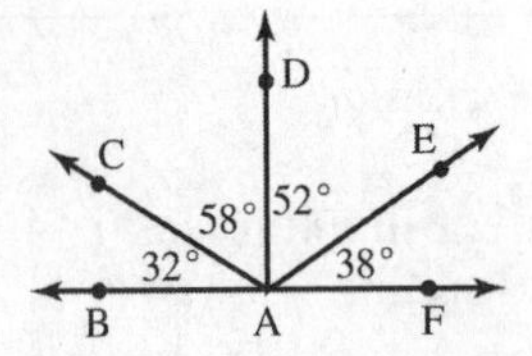

12. **12.** ________

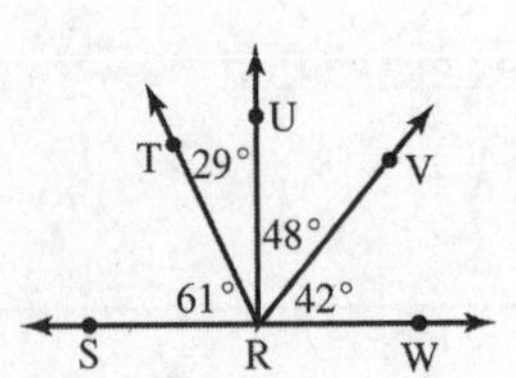

Identify each pair of supplementary angles.

13. **13.** ________

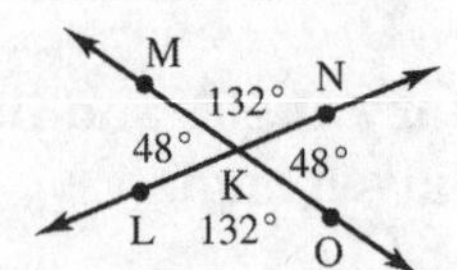

14. **14.** ________

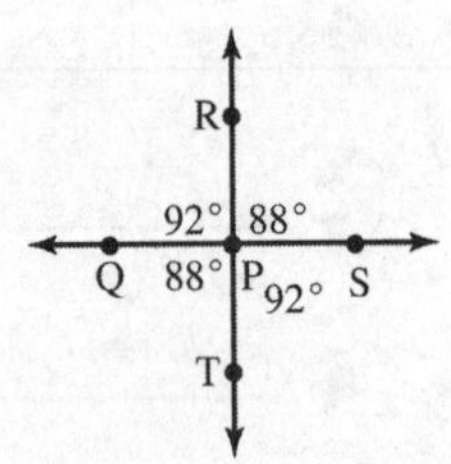

Objective 2 Identify congruent angles and vertical angles and use this knowledge to find the measures of angles.

In each of the following, identify the angles that are congruent.

15. **15.** ________

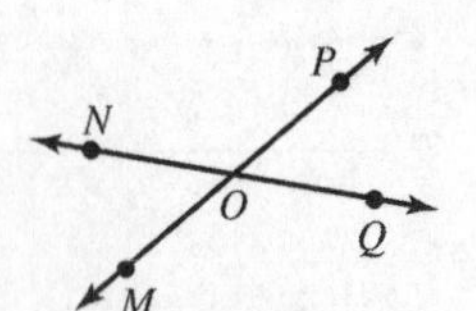

Name: Date:

Instructor: Section:

16. 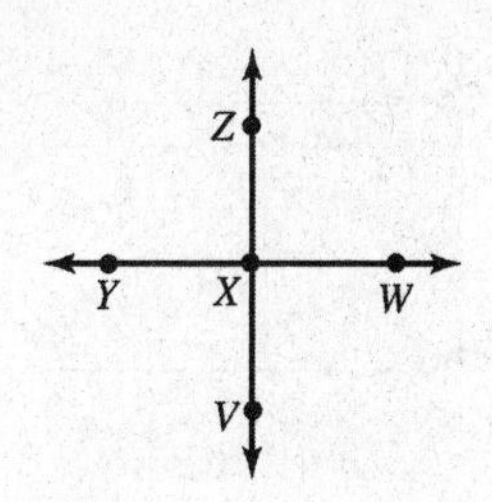

16. ______________

17. 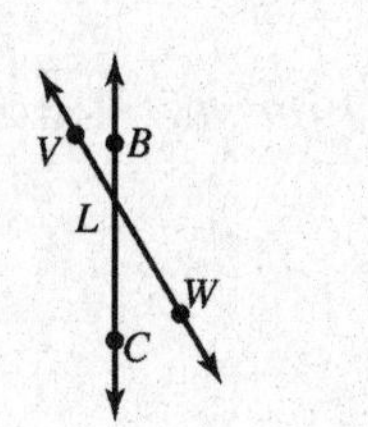

17. ______________

18. 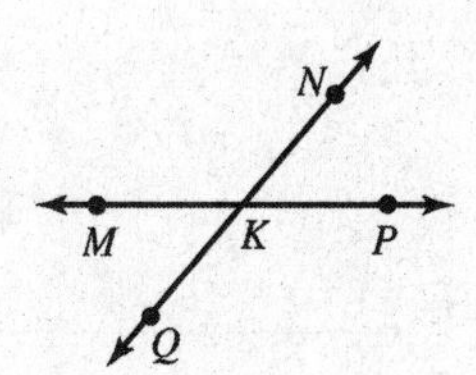

18. ______________

19. 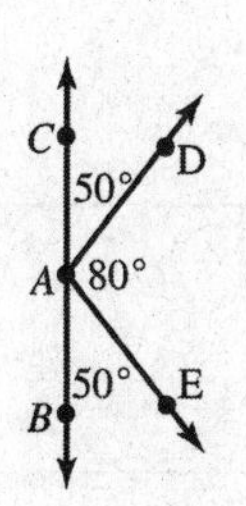

19. ______________

20. 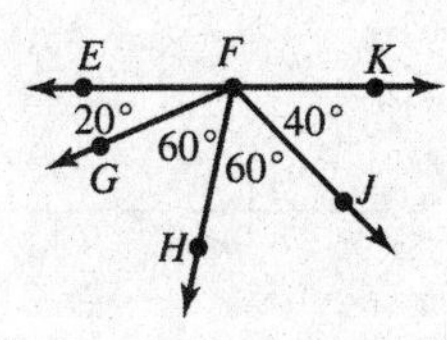

20. ______________

In the figure below, $\angle AGF$ measures 33° and $\angle BGC$ measures 105°. Find the measures of the indicated angles.

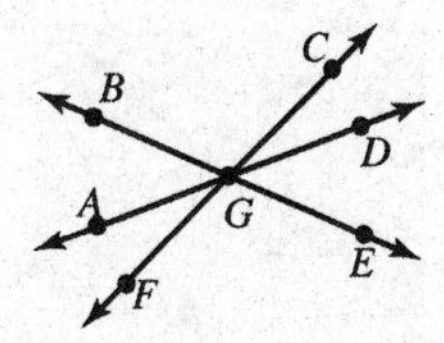

21. $\angle CGD$

21. ______________

22. $\angle EGF$

22. ______________

Name: Date:
Instructor: Section:

23. $\angle DGE$ **23.** ________________

24. $\angle BGA$ **24.** ________________

In the figure below, $\angle APB$ measures 93° and $\angle BPC$ measures 37°. Find the measures of the indicated angles.

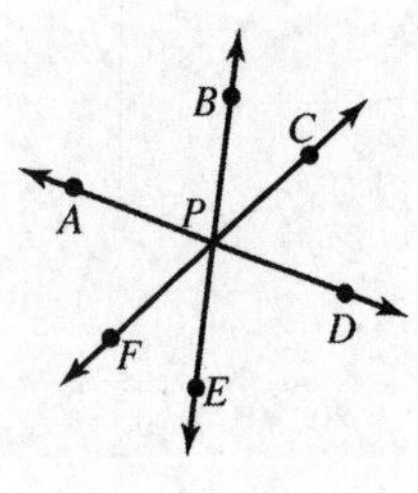

25. $\angle CPD$ **25.** ________________

26. $\angle DPE$ **26.** ________________

27. $\angle EPF$ **27.** ________________

28. $\angle FPA$ **28.** ________________

In the figure below, $\angle ABE$ measures 73° and $\angle FEB$ measures 107°. Find the measures of the indicated angles.

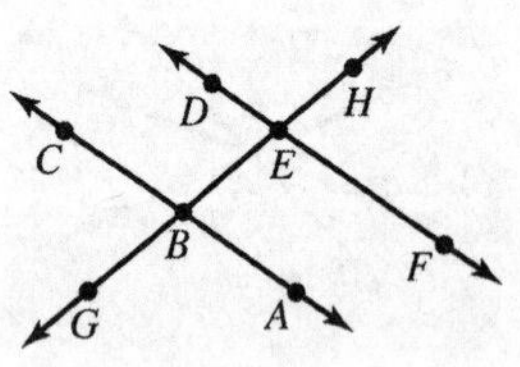

29. $\angle CBG$ **29.** ________________

30. $\angle DEB$ **30.** ________________

Name: Date:
Instructor: Section:

Chapter 8 GEOMETRY

8.3 Rectangles and Squares

Learning Objectives
1 Find the perimeter and area of a rectangle.
2 Find the perimeter and area of a square.
3 Find the perimeter and area of a composite figure.

Key Terms

Use the vocabulary terms listed below to complete each statement in exercises 1–4.

perimeter **area** **rectangle** **square**

1. The number of square units in a region is called the ________________ of the region.

2. A four-sided figure with four right angles is called a ____________________.

3. The distance around the outside edges of a figure is called the ____________________ of the figure.

4. A rectangle with four equal sides is called a ____________________.

Objective 1 Find the perimeter and area of a rectangle.

Find the perimeter and area of each rectangle.

1. 4 centimeters by 8 centimeters — 1. ________________

2. 17 inches by 12 inches — 2. ________________

3. 1 centimeter by 17 centimeters — 3. ________________

4. 14.5 meters by 3.2 meters — 4. ________________

5. $4\frac{1}{2}$ yards by $6\frac{1}{2}$ yards — 5. ________________

6. 87.2 feet by 33 feet — 6. ________________

7. 37.4 centimeters by 103.2 centimeters — 7. ________________

Name: Date:
Instructor: Section:

Solve each application problem.

8. A picture frame measures 20 inches by 30 inches. Find the perimeter and area of the frame. **8.** ________

9. A lot is 114 feet by 212 feet. County rules require that nothing be built on land within 12 feet of any edge of the lot. Find the area on which you cannot build. **9.** ________

10. A room is 14 yards by 18 yards. Find the cost to carpet this room if carpet costs $23 per square yard. **10.** ________

Objective 2 Find the perimeter and area of a square.

Find the perimeter and area of each square with the given side.

11. 9 meters **11.** ________

12. 9.2 yards **12.** ________

13. 7.8 feet **13.** ________

14. 13 feet **14.** ________

15. $1\frac{2}{5}$ inches **15.** ________

16. 8.2 km **16.** ________

17. 3.1 cm **17.** ________

18. 7.4 inches **18.** ________

Name: Date:
Instructor: Section:

19. $4\frac{2}{3}$ miles **19.** ______________

20. 21 m **20.** ______________

Objective 3 Find the perimeter and area of a composite figure.

Find the perimeter and area of each figure. All angles that appear to be right angles are, in fact, right angles.

21. **21.** ______________

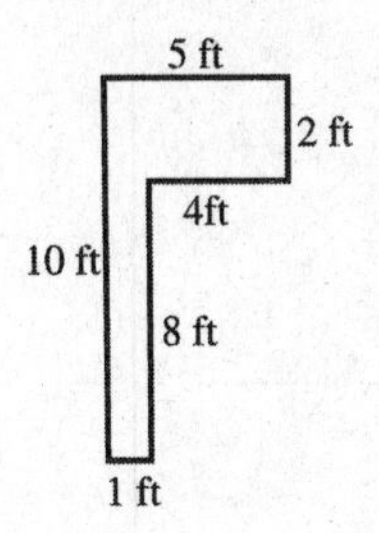

22. **22.** ______________

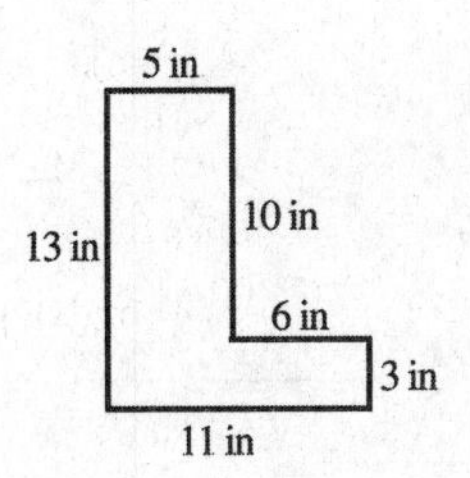

23. **23.** ______________

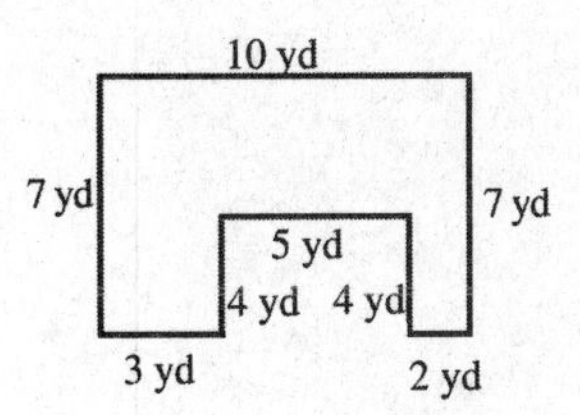

24. **24.** ______________

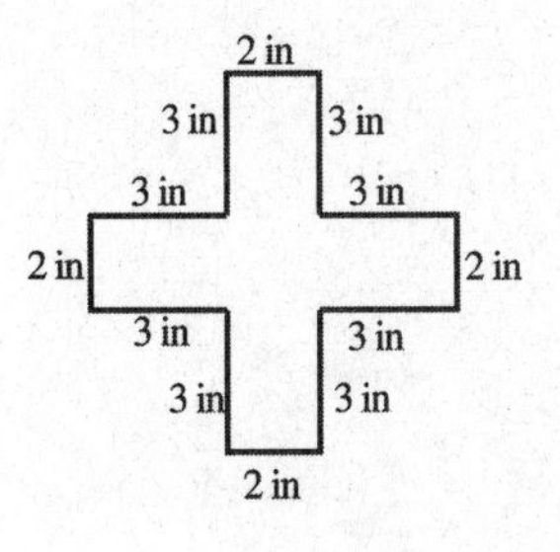

25. **25.** ______________

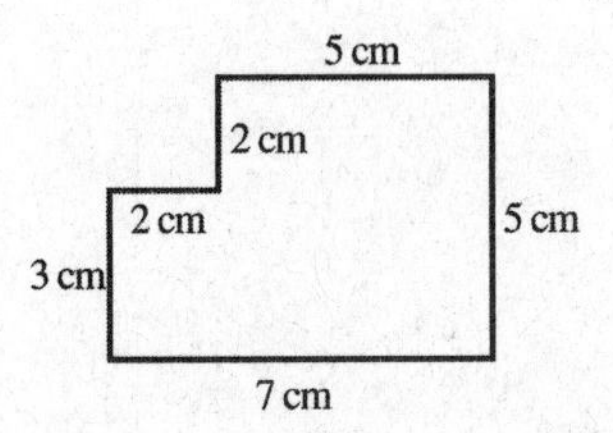

Name: Date:
Instructor: Section:

26.

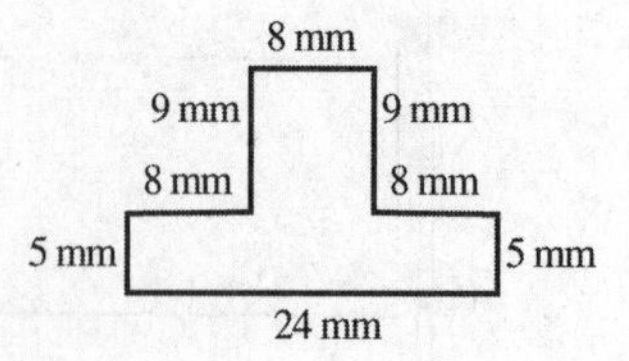

26. ______________

27.

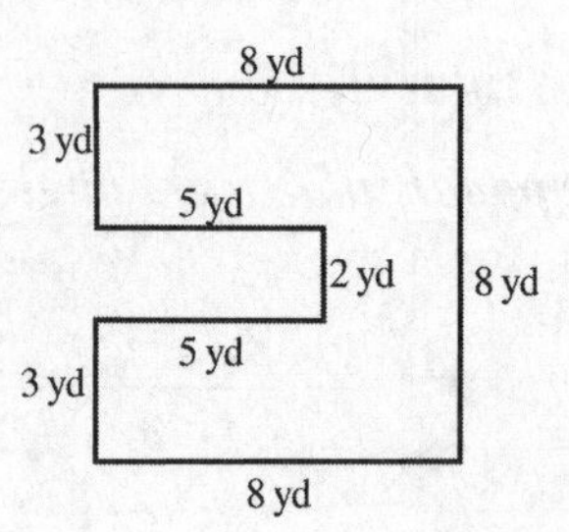

27. ______________

28.

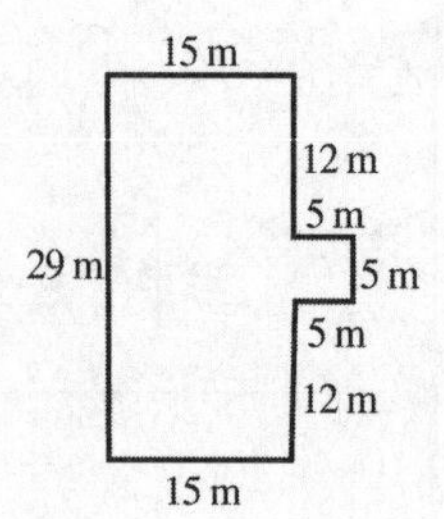

28. ______________

29.

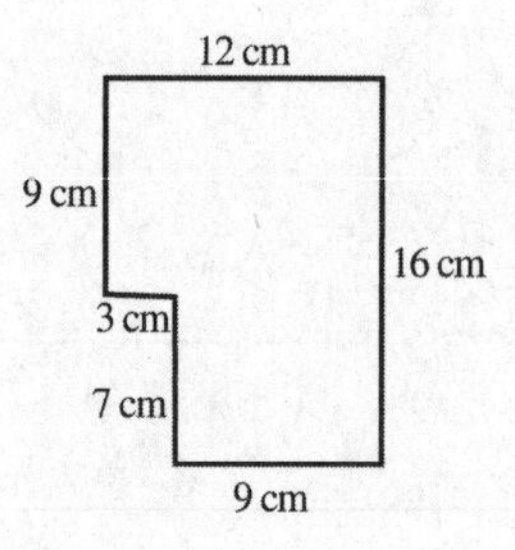

29. ______________

30.

10 ft
20 ft
18 ft
10 ft
40 ft
14 ft
32 ft
44 ft

30. ______________

Name: Date:
Instructor: Section:

Chapter 8 GEOMETRY

8.4 Parallelograms and Trapezoids

Learning Objectives	
1	Find the perimeter and area of a parallelogram.
2	Find the perimeter and area of a trapezoid.

Key Terms

Use the vocabulary terms listed below to complete each statement in exercises 1–4.

perimeter **area** **parallelogram** **trapezoid**

1. A ______________________ is a four-sided figure with both pairs of opposite sides parallel and equal in length.

2. A ______________________ is a four-sided figure with exactly one pair of parallel sides.

3. The formula $P = 2 \cdot l + 2 \cdot w$ is the formula for the ______________________ of a rectangle.

4. Square the length of a side of a square to find the ______________________ of a square.

Objective 1 Find the perimeter and area of a parallelogram.

Find the perimeter of each parallelogram.

1.

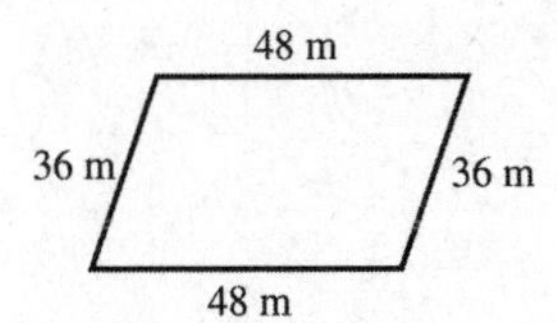

1. ________________

2.

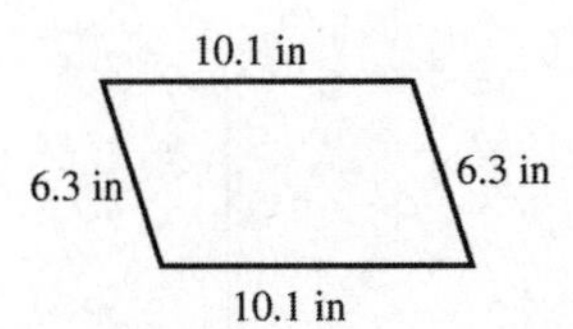

2. ________________

3.

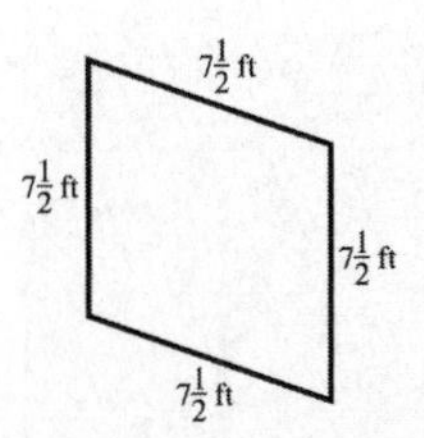

3. ________________

Name: Date:
Instructor: Section:

4.

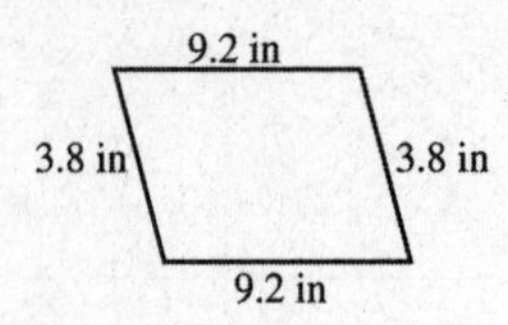

4. ______________

5.

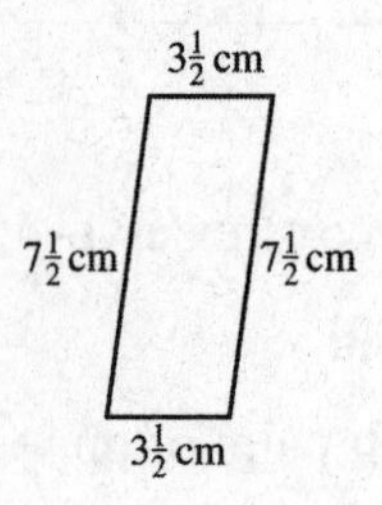

5. ______________

Find the area of each parallelogram.

6.

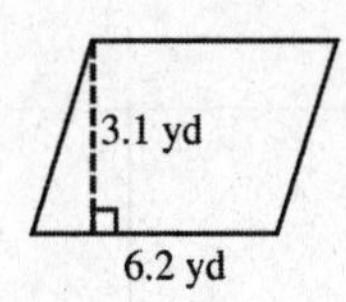

6. ______________

7.

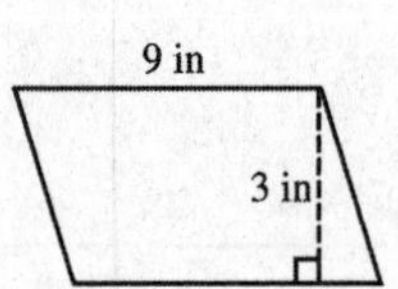

7. ______________

8.

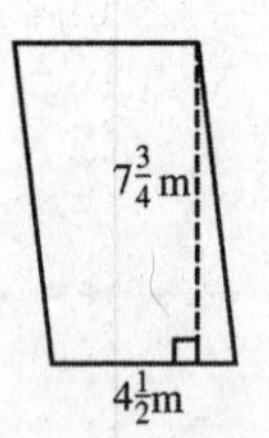

8. ______________

9.

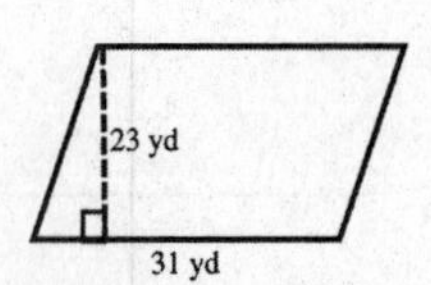

9. ______________

Name: Date:
Instructor: Section:

10. **10.** ______________

9.8 m

12.6 m

11. **11.** ______________

$2\frac{1}{2}$ m

$4\frac{1}{2}$ m

Solve each application problem.

12. A parallelogram has a height of 3.2 meters and a base of 4.6 meters. Find the area. **12.** ______________

13. A parallelogram has a height of $15\frac{1}{2}$ feet and a base of 20 feet. Find the area. **13.** ______________

14. A swimming pool is in the shape of a parallelogram with a height of 9.6 meters and base of 12 meters. Find the cost of a solar pool cover that sells for \$5.10 per square meter. **14.** ______________

15. An auditorium stage has a hardwood floor that is shaped like a parallelogram, having a height of 30 feet and a base of 40 feet. If a company charges \$0.65 per square foot to refinish floors, find the cost of refinishing the stage floor. **15.** ______________

Name: Date:
Instructor: Section:

Objective 2 Find the perimeter and area of a trapezoid.

Find the perimeter of each figure.

16.

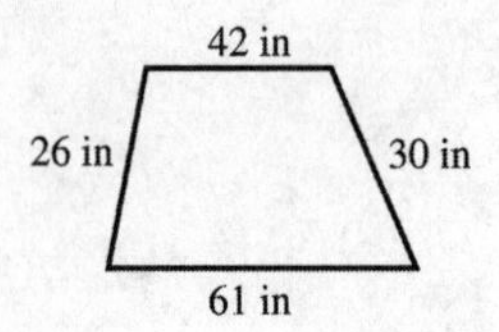

16. ______________

17.

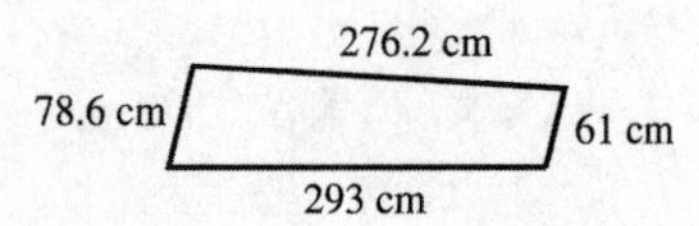

17. ______________

18.

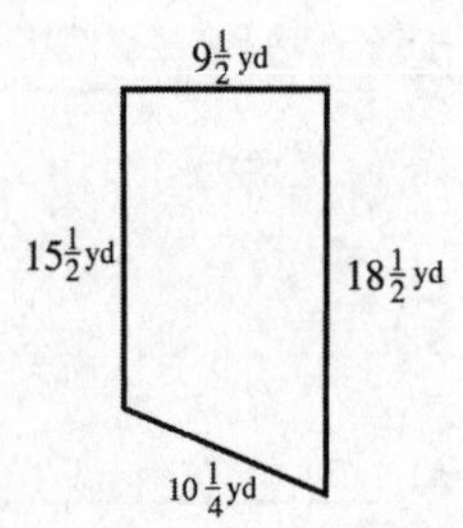

18. ______________

Find the area of each figure.

19.

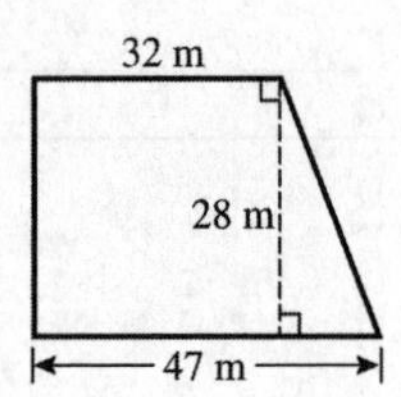

19. ______________

20.

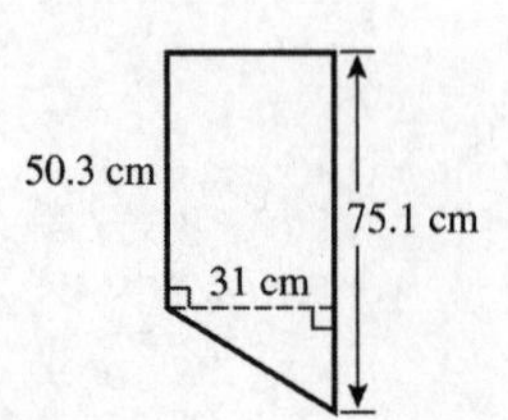

20. ______________

Name: Date:
Instructor: Section:

21.

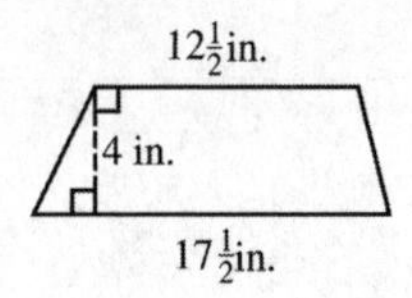

21. ________________

22.

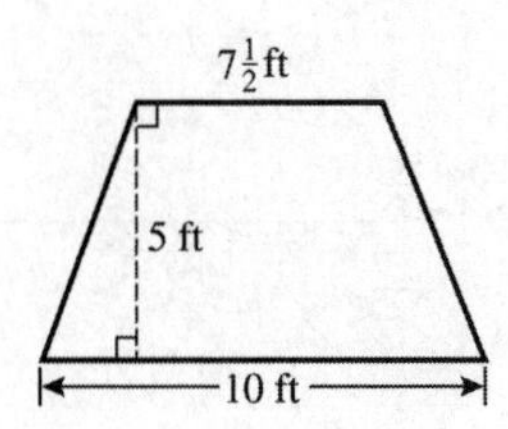

22. ________________

23.

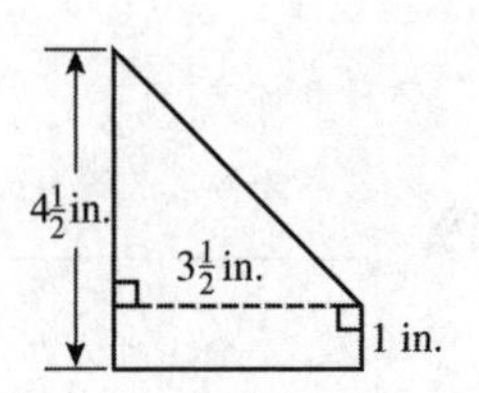

23. ________________

24.

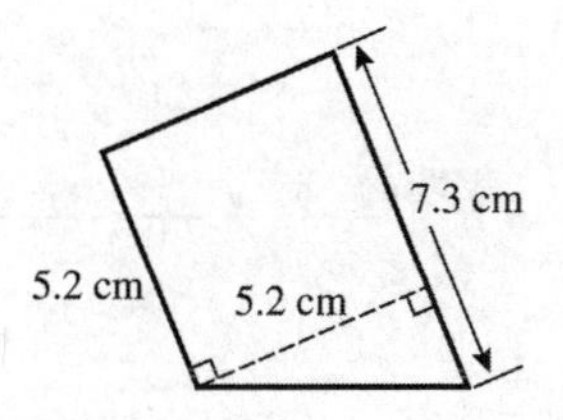

24. ________________

25.

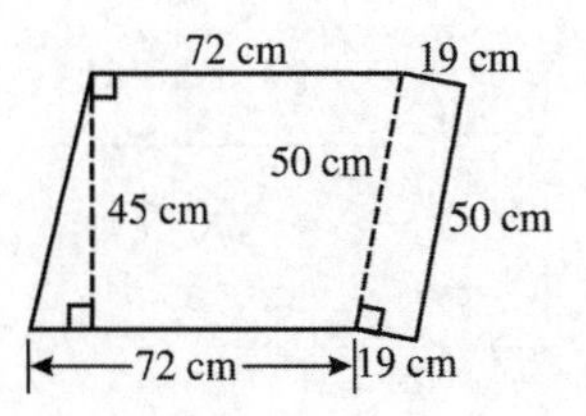

25. ________________

Name: Date:
Instructor: Section:

26. 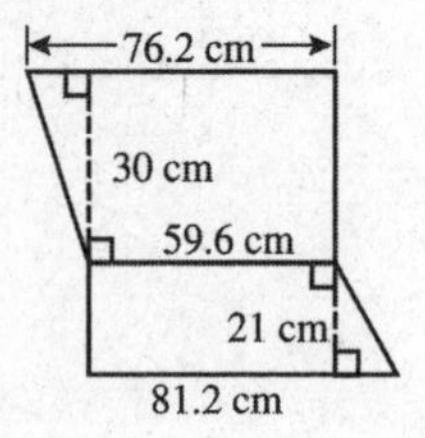

26. ________________

27. 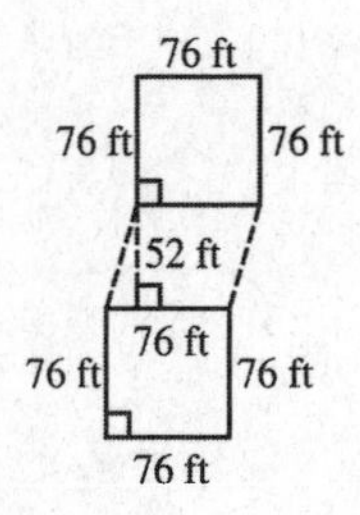

27. ________________

Solve each application problem.

28. The lobby in a resort hotel is in the shape of a trapezoid. The height of the trapezoid is 52 feet and the bases are 47 feet and 59 feet. Carpet that costs $2.75 per square foot is to be laid in the lobby. Find the cost of the carpet.

28. ________________

29. The backyard of a new home is shaped like a trapezoid, having a height of 35 feet and bases of 90 feet and 110 feet. Find the cost of planting a lawn in the yard if the landscaper charges $0.20 per square foot.

29. ________________

30. A hot tub is in the shape shown. Find the cost of a cover for the hot tub at a cost of $9.70 per square foot. Angles that appear to be right angles are indeed right angles.

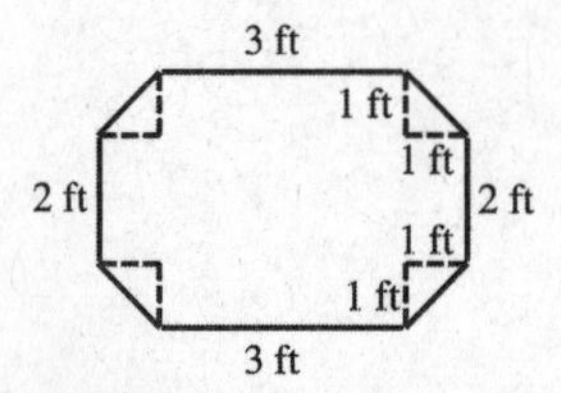

30. ________________

Name: Date:
Instructor: Section:

Chapter 8 GEOMETRY

8.5 Triangles

Learning Objectives	
1	Find the perimeter of a triangle.
2	Find the area of a triangle.
3	Given the measures of two angles in a triangle, find the measure of the third angle.

Key Terms

Use the vocabulary terms listed below, along with the figure, to complete each statement in exercises 1–3.

base **height** **triangle**

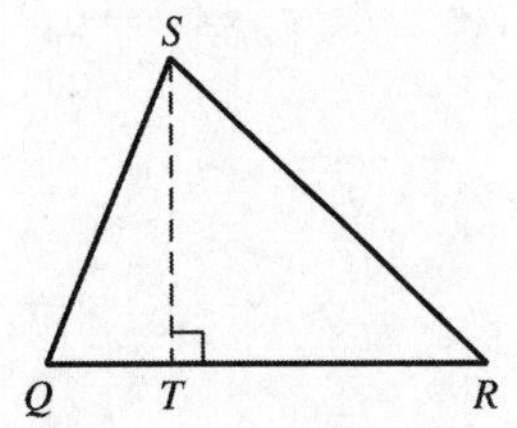

1. A figure with exactly three sides is called a ______________________.

2. In the figure, $\overline{QR}$ is the ______________________ of ⊔ *QRS*.

3. In the figure, $\overline{ST}$ is the ______________________ of ⊔ *QRS*.

Objective 1 Find the perimeter of a triangle.

Find the perimeter of each triangle.

1. 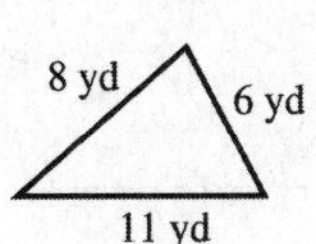

1. ______________

2. 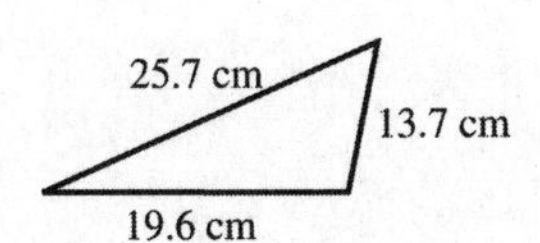

2. ______________

Name: Date:
Instructor: Section:

3.

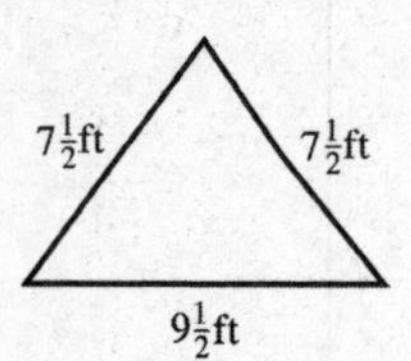

3. ______________

4.

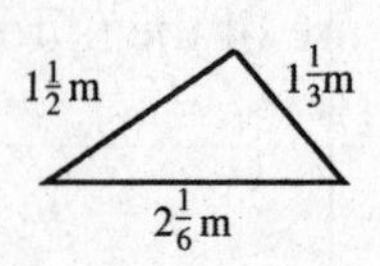

4. ______________

5.

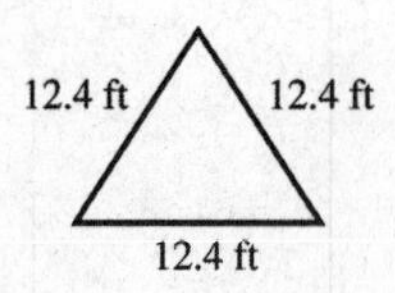

5. ______________

6.

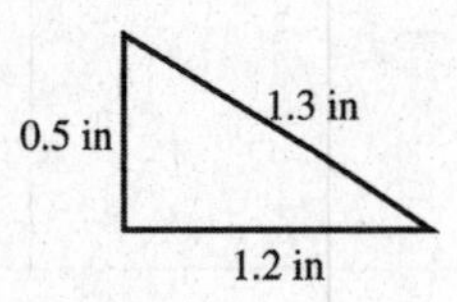

6. ______________

7. A triangle with sides $2\frac{1}{2}$ feet, 3 feet, and $5\frac{1}{4}$ feet

7. ______________

8. A triangle with two equal sides of 3.6 centimeters and the third side 4.1 centimeters

8. ______________

9. A triangle with three equal sides each 5.9 meters

9. ______________

Name: Date:
Instructor: Section:

10. A triangle with sides $13\frac{1}{8}$ inches, $11\frac{3}{4}$ inches, and $14\frac{1}{2}$ inches.

10. ____________

Objective 2 Find the perimeter of a triangle.

Find the area of each triangle.

11.

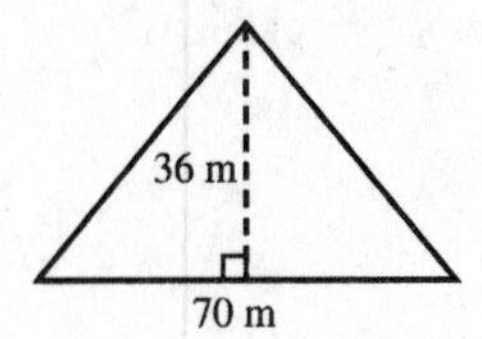

11. ____________

12.

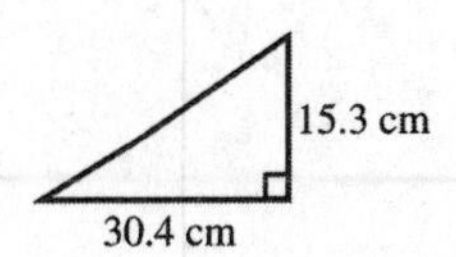

12. ____________

13.

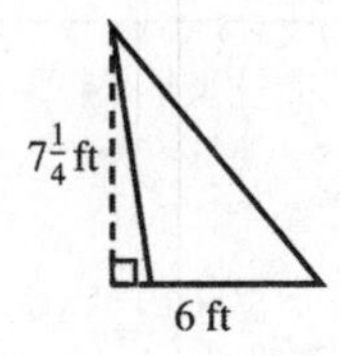

13. ____________

14.

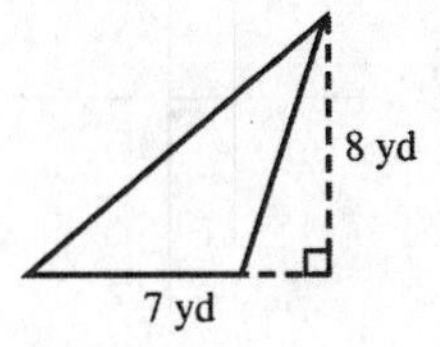

14. ____________

15.

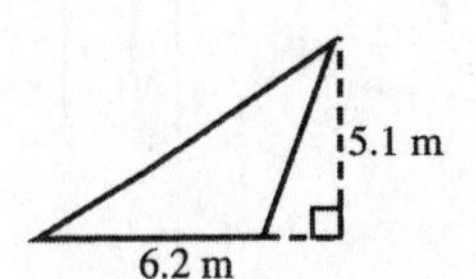

15. ____________

Name: Date:
Instructor: Section:

16.

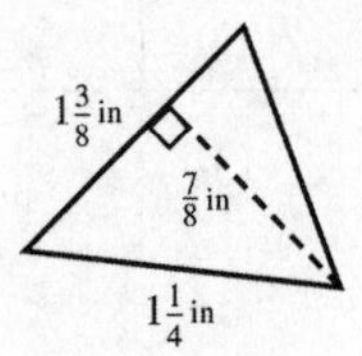

16. ______________

Find the shaded area in each figure.

17.

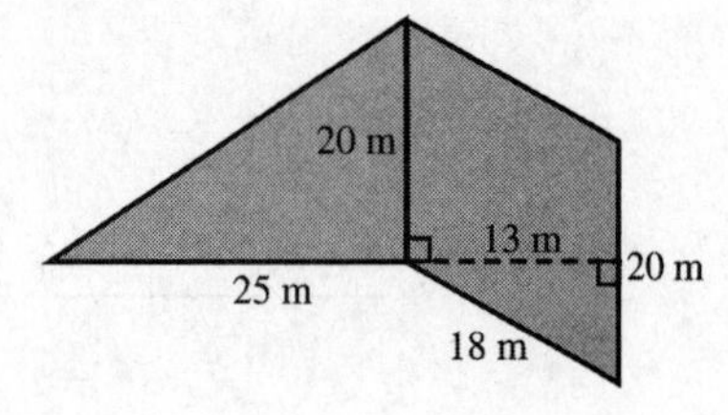

17. ______________

18.

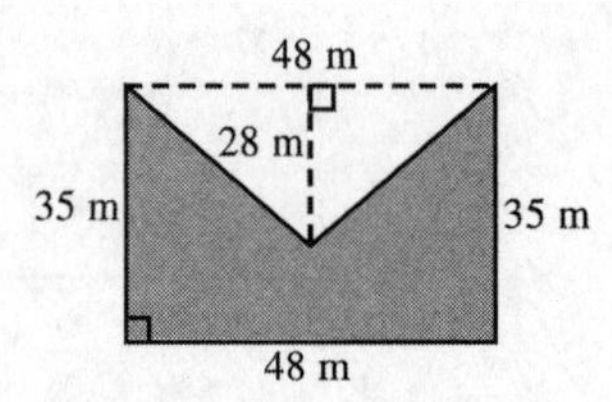

18. ______________

19.

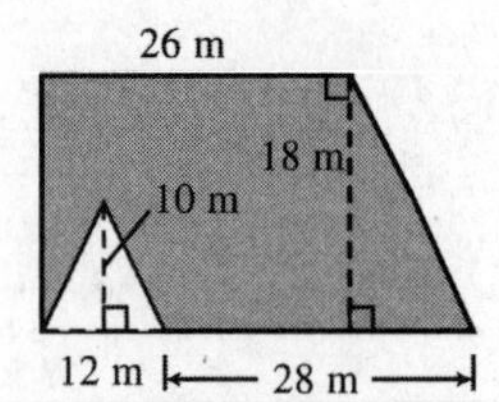

19. ______________

20.

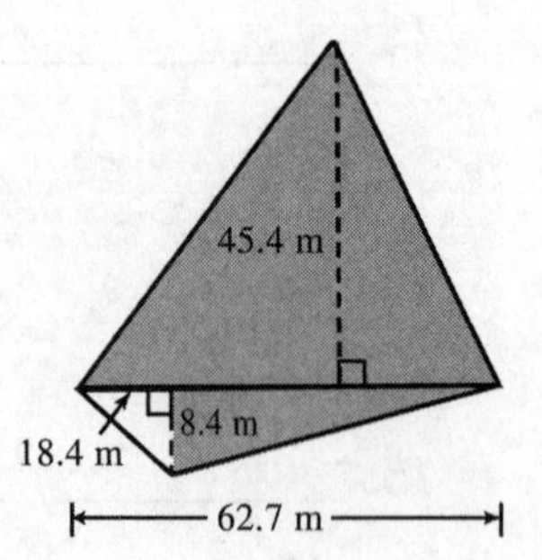

20. ______________

Name: Date:
Instructor: Section:

21. 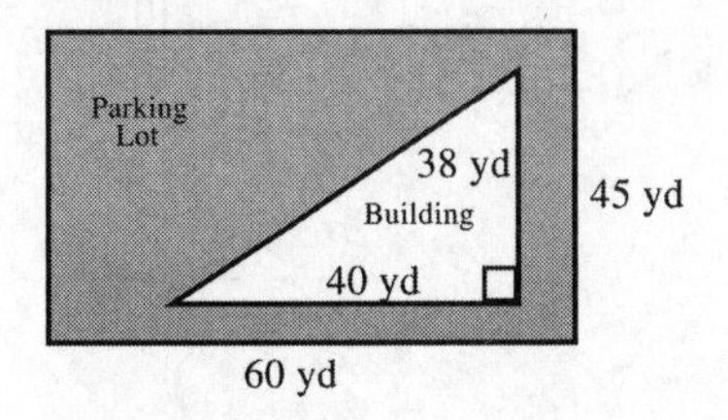

21. ________________

22. If a painter charged $4.06 per square meter to paint the front of the house shaded below, how much would he charge? All angles that appear to be right angles are right angles.

22. ________________

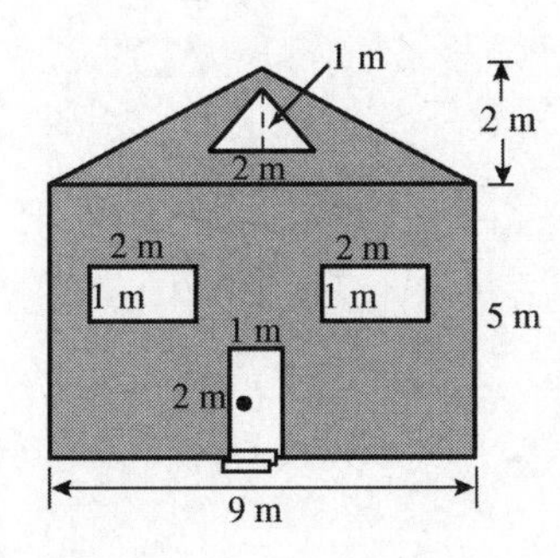

Objective 3 Given the measures of two angles in a triangle, find the measure of the third angle.

The measures of two angles of a triangle are given. Find the measure of the third angle.

23. 100°, 63° 23. ________________

24. 60°, 60° 24. ________________

25. 37°, 62° 25. ________________

26. 49°, 72° 26. ________________

27. 51°, 78° 27. ________________

Name: Date:
Instructor: Section:

28. 87°, 13° **28.** ________________

29. 90°, 45° **29.** ________________

30. 76°, 76° **30.** ________________

Name: Date:
Instructor: Section:

Chapter 8 GEOMETRY

8.6 Circles

Learning Objectives	
1	Find the radius and diameter of a circle.
2	Find the circumference of a circle.
3	Find the area of a circle.
4	Become familiar with Latin and Greek prefixes used in math terminology.

Key Terms

Use the vocabulary terms listed below to complete each statement in exercises 1–5.

circle **radius** **diameter** **circumference** **π (pi)**

1. The ____________________ is the distance from the center of a circle to any point on the circle.

2. The ____________________ of a circle is the distance around the circle.

3. A figure whose points lie the same distance from a fixed center point is called a ____________________.

4. The ratio of the circumference to the diameter of any circle equals ____________.

5. The ________________________ of a circle is a segment connecting two points on a circle and passing through the center.

Objective 1 Find the radius and diameter of a circle.

Find the diameter or radius in each circle.

1.

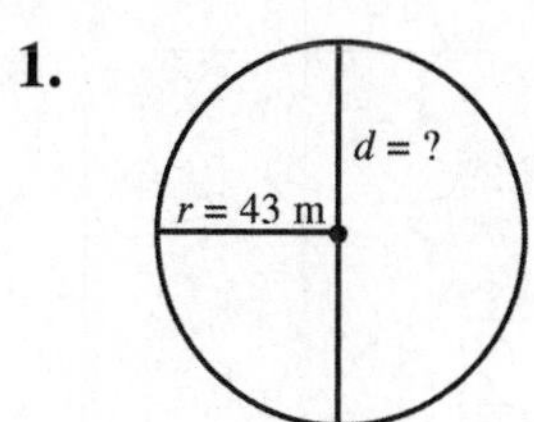

1. ________________

2.

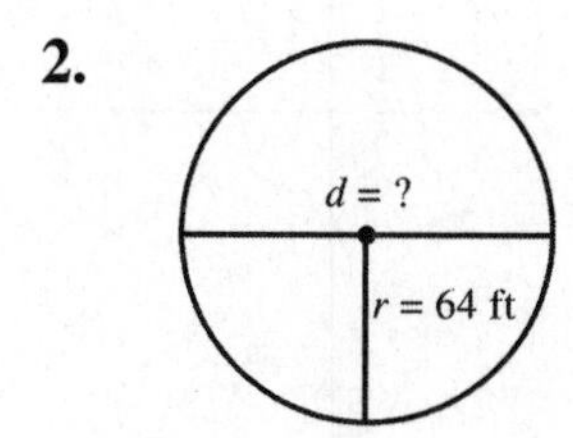

2. ________________

Name: Date:
Instructor: Section:

3. 3. ______________

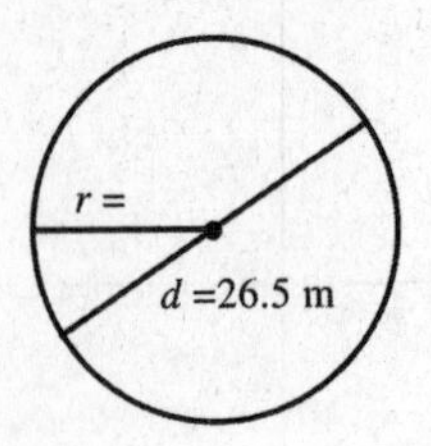

4. 4. ______________

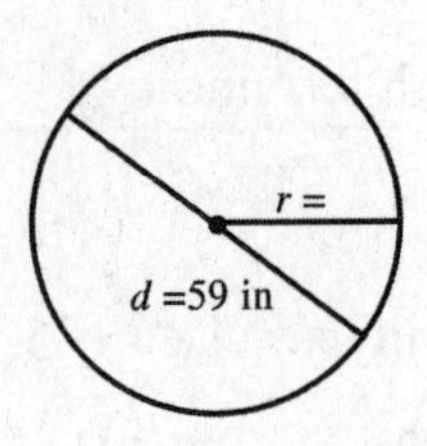

5. The diameter of a circle is 8 feet. Find its radius. 5. ______________

6. The radius of a circle is 2.7 centimeters. Find its diameter. 6. ______________

7. The diameter of a circle is $12\frac{1}{2}$ yards. Find its radius 7. ______________

Objective 2 Find the circumference of a circle.

Find the circumference of each circle. Use 3.14 as an approximation for π. Round each answer to the nearest tenth.

8. 8. ______________

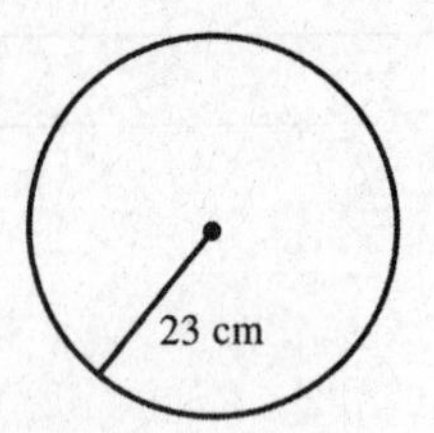

9. 9. ______________

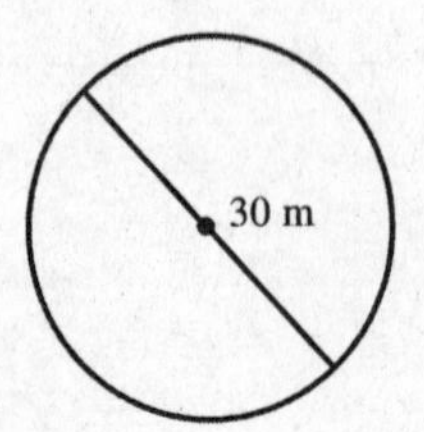

Name: Date:
Instructor: Section:

10. A circle with a diameter of $4\frac{3}{4}$ inches **10.** ________

11. A circle with a radius of 4.5 yards **11.** ________

12. A circle with a radius of $\frac{3}{4}$ mile **12.** ________

Solve each application problem.

13. How far does a point on the tread of a tire move in one turn if the diameter of the tire is 60 centimeters? **13.** ________

14. If you swing a ball held at the end of a string 3 meters long, how far will the ball travel on each turn? **14.** ________

Objective 3 Find the area of a circle.

Find the area of each circle. Use 3.14 as an approximation for π. Round each answer to the nearest tenth.

15. **15.** ________

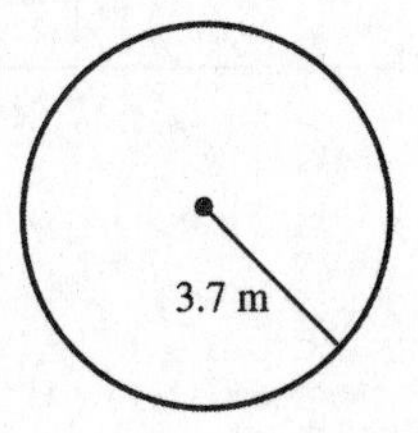

Name: Date:
Instructor: Section:

16. 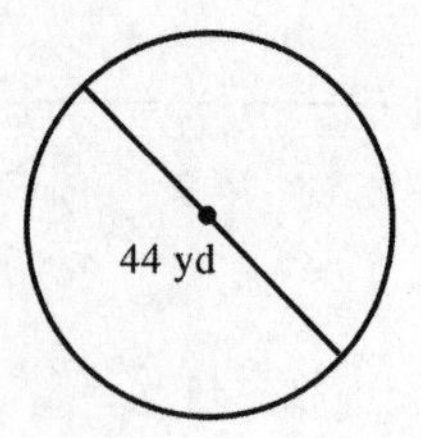

16. ______________

17. A circle with diameter of $5\frac{1}{3}$ yards **17.** ______________

18. A circle with diameter of 9.8 centimeters **18.** ______________

Find the area of the shaded region. Use 3.14 as an approximation for π. Round each answer to the nearest tenth.

19. 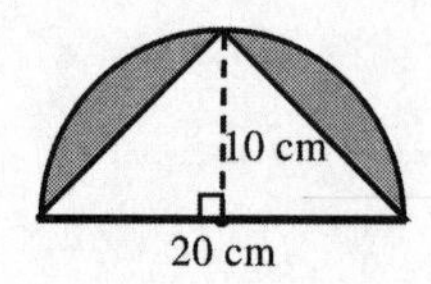

19. ______________

20. 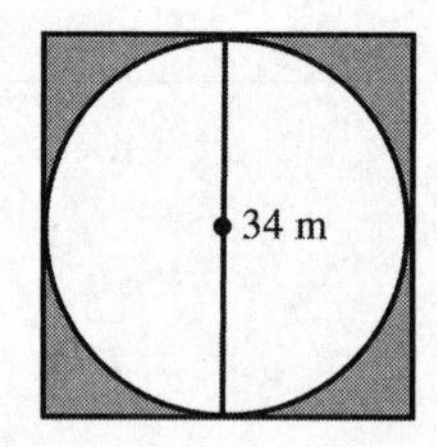

20. ______________

21. 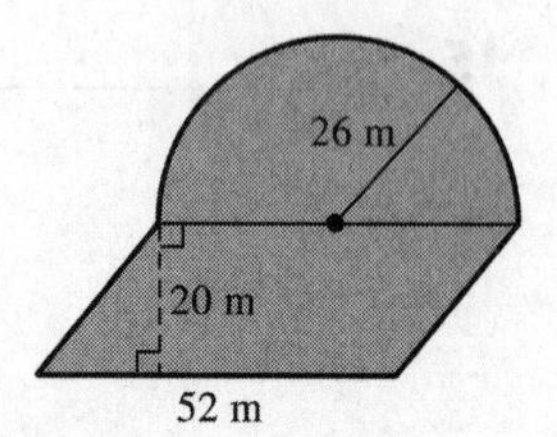

21. ______________

Name: Date:
Instructor: Section:

22.

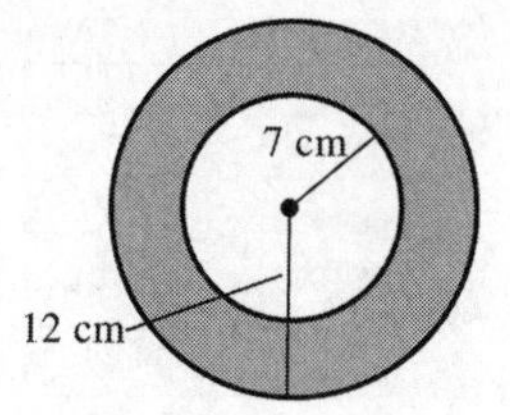

22. ______________

23.

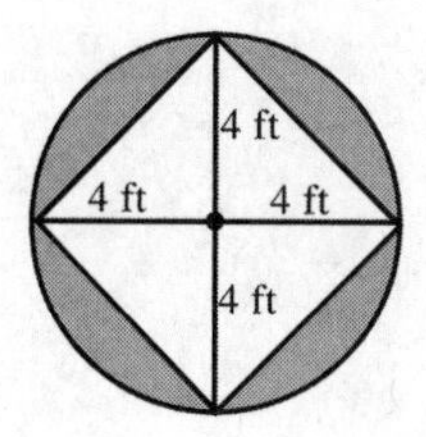

23. ______________

Solve each application problem.

24. Find the area of a circular pond that has a diameter of 12.6 meters.

24. ______________

25. Find the cost of sod, at \$1.80 per square foot, for the following playing field. Round the answer to the nearest cent.

25. ______________

Objective 4 Become familiar with Latin and Greek prefixes used in math terminology.

Write one math term and one nonmathematical term that use each prefix listed below. (Answers will vary.)

26. *tri-* (three)

26. ______________

Name: Date:
Instructor: Section:

27. *poly*-(many) **27.** ________________

28. *milli*-(thousand) **28.** ________________

29. *oct*-(eight) **29.** ________________

30. *uni*-(one) **30.** ________________

Name: Date:
Instructor: Section:

Chapter 8 GEOMETRY

8.7 Volume

Learning Objectives	
Find the volume of a	
1	rectangular solid.
2	sphere.
3	cylinder.
4	cone and pyramid.

Key Terms

Use the vocabulary terms listed below to complete each statement in exercises 1–6.

volume **rectangular solid** **sphere** **cylinder** **cone** **pyramid**

1. A ______________________________ is a box-like solid figure.

2. A solid figure with two congruent, parallel, circular bases is a ________________.

3. A ____________________ is a ball-like solid figure.

4. ____________________ is a measure of the space inside a solid shape.

5. A solid figure whose base is a square or a rectangle and whose faces (sides) are triangles is called a ____________________.

6. A solid figure with only one base, and that base is a circle, is called a ___________.

Objective 1 Find the volume of a rectangular solid.

Find the volume of each rectangular solid. Round answers to the nearest tenth, if necessary.

1.

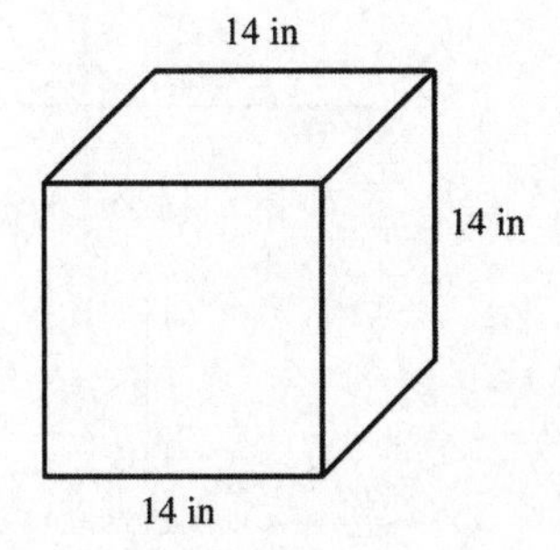

1. ______________

Name: Date:
Instructor: Section:

2.

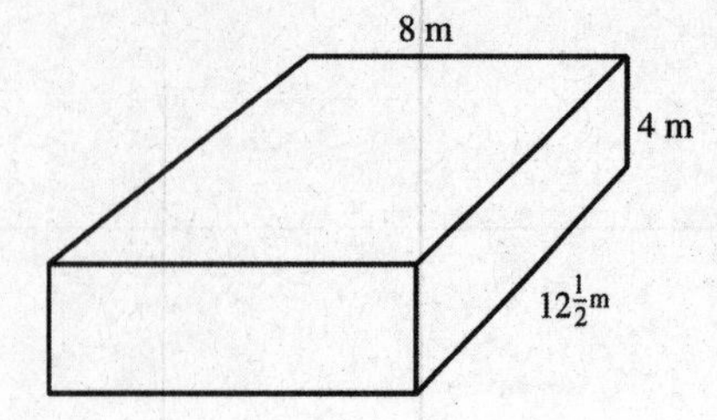

2. ______________

3.

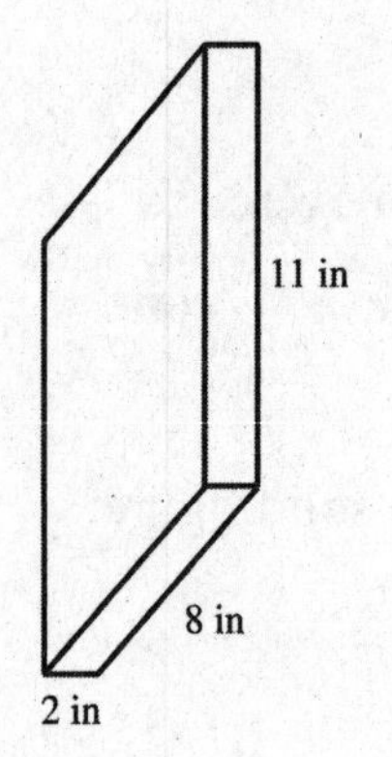

3. ______________

4.

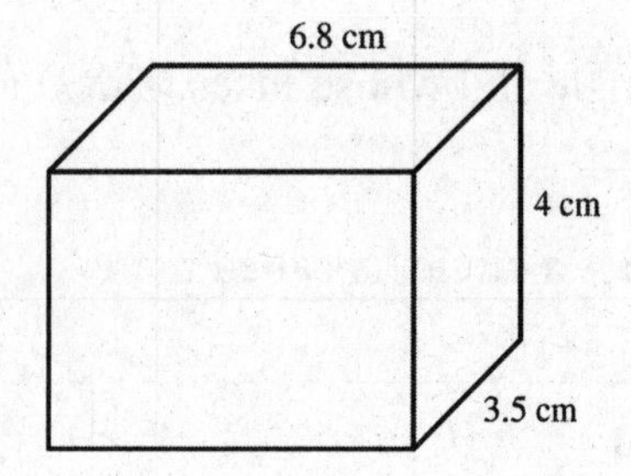

4. ______________

5.

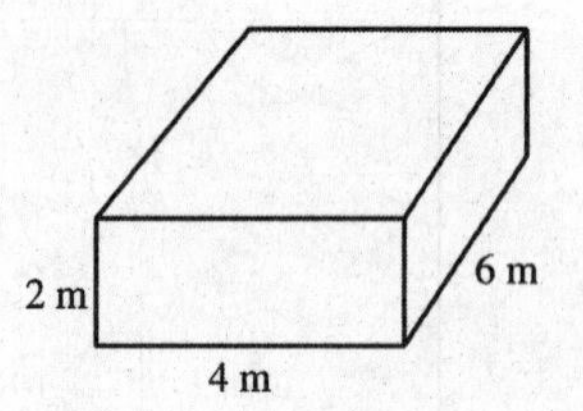

5. ______________

Name: Date:
Instructor: Section:

Find the volume of each solid.

6. **6.** ________________

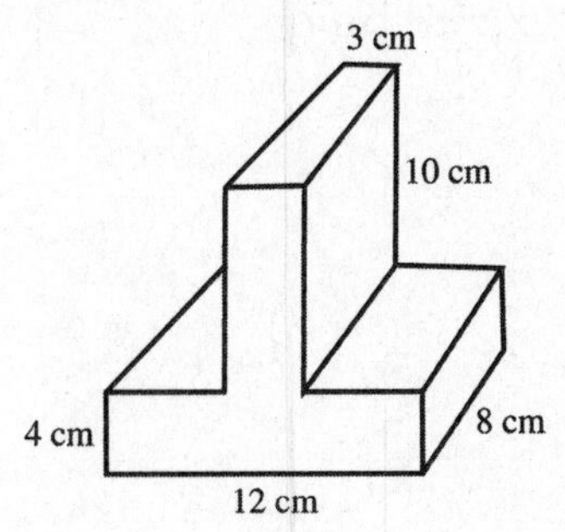

7. **7.** ________________

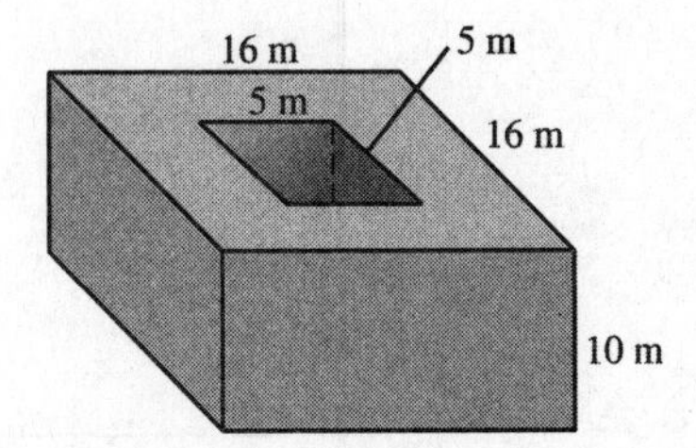

(Notice the square hole that goes through the center of the shape.)

Objective 2 Find the volume of a sphere.

Find the volume of each sphere or hemisphere. Use 3.14 as an approximation for π. Round answers to the nearest tenth, if necessary.

8. **8.** ________________

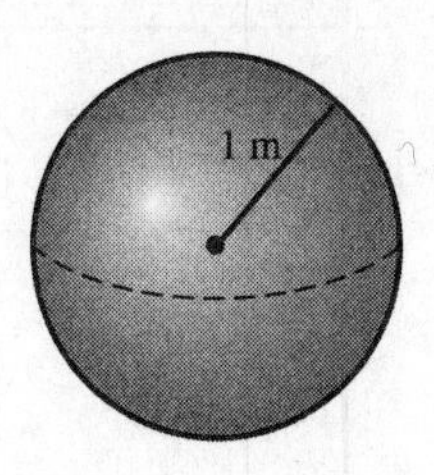

9. **9.** ________________

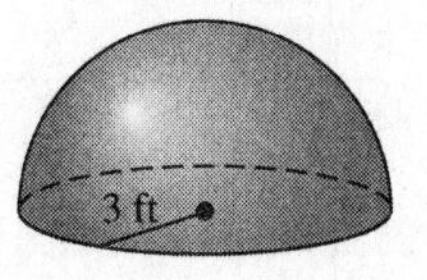

Name: Date:
Instructor: Section:

10.

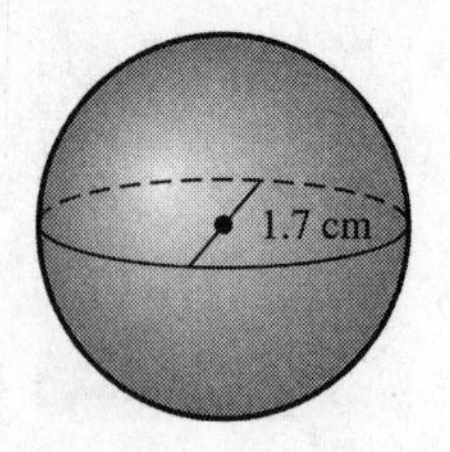

10. ______________

11.

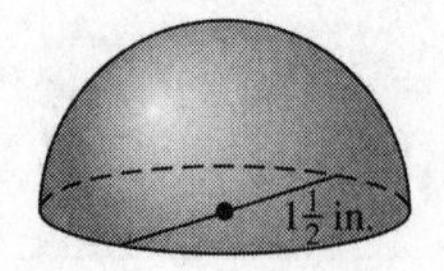

11. ______________

12. A sphere with a diameter of $3\frac{1}{4}$ inches. **12.** ______________

13. A hemisphere with a radius of 11.6 feet. **13.** ______________

14. A sphere with a radius of 6.8 cm **14.** ______________

Objective 3 Find the volume of a cylinder.

Find the volume of each figure. Use 3.14 as an approximation for π. Round answers to the nearest tenth, if necessary.

15.

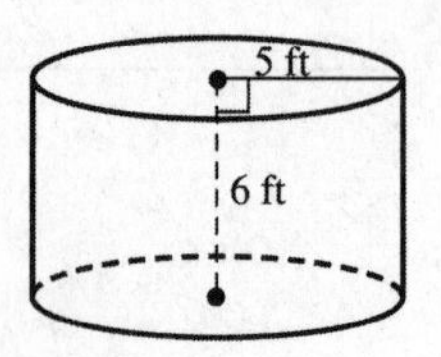

15. ______________

Name: Date:
Instructor: Section:

16.

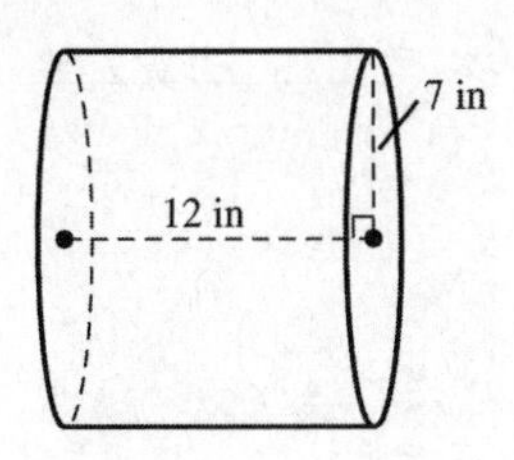

16. ______________

17.

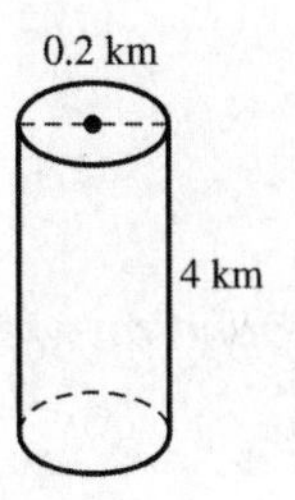

17. ______________

18. A coffee can, radius 6 centimeters and height 16 centimeters

18. ______________

19. An oil can, diameter 8 centimeters and height 13.5 centimeters

19. ______________

20. A cardboard mailing tube, diameter 5 centimeters and height 25 centimeters

20. ______________

21.

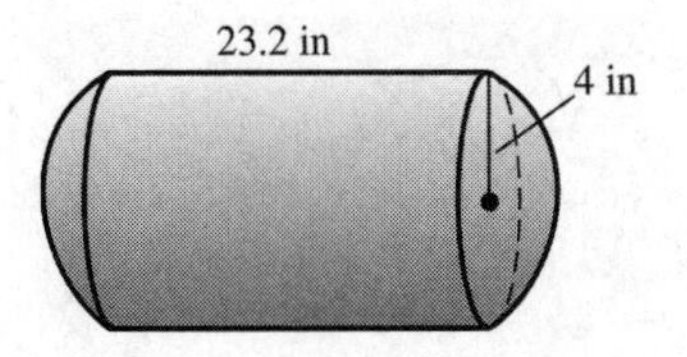

21. ______________

Name: Date:
Instructor: Section:

22.

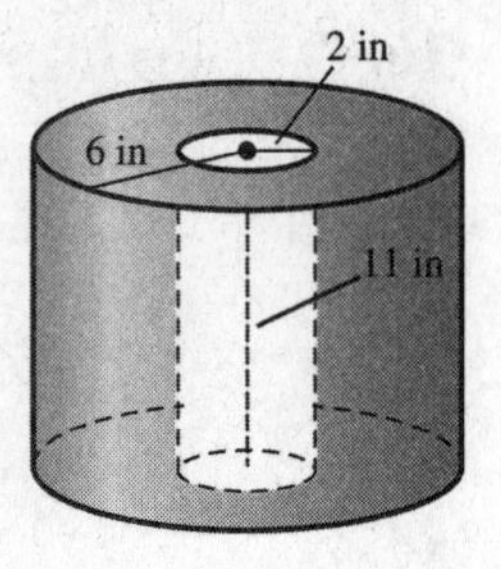

22. ______________

Objective 4 Find the volume of a cone and a pyramid.

Find the volume of each figure. Use 3.14 as an approximation for π. Round answers to the nearest tenth, if necessary.

23.

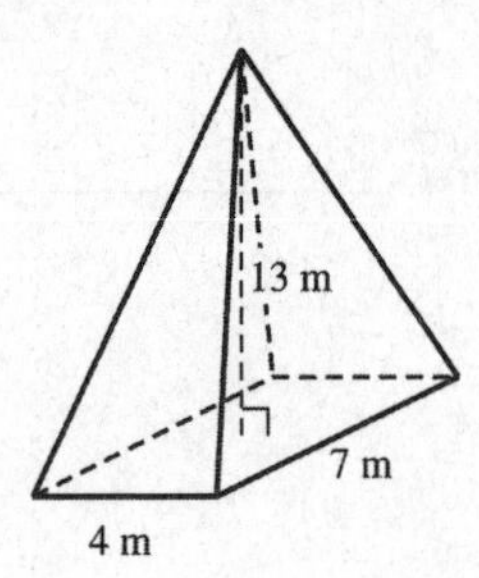

23. ______________

24.

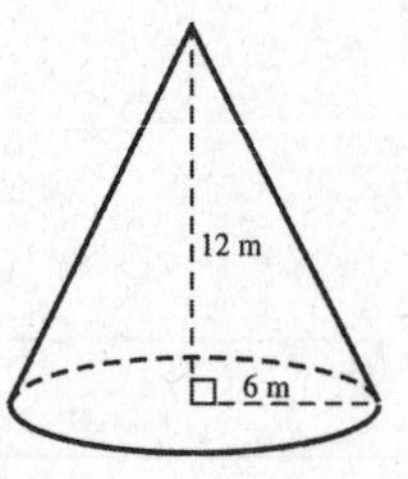

24. ______________

25.

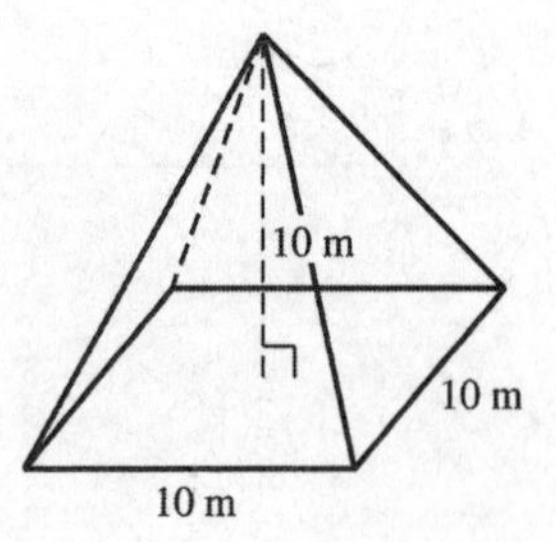

25. ______________

26.

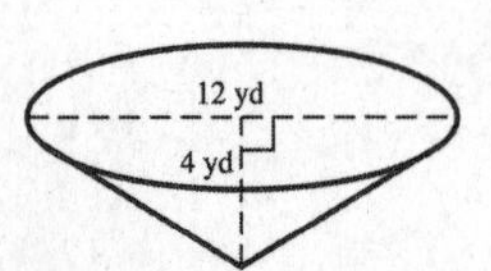

26. ______________

Name: Date:
Instructor: Section:

27.

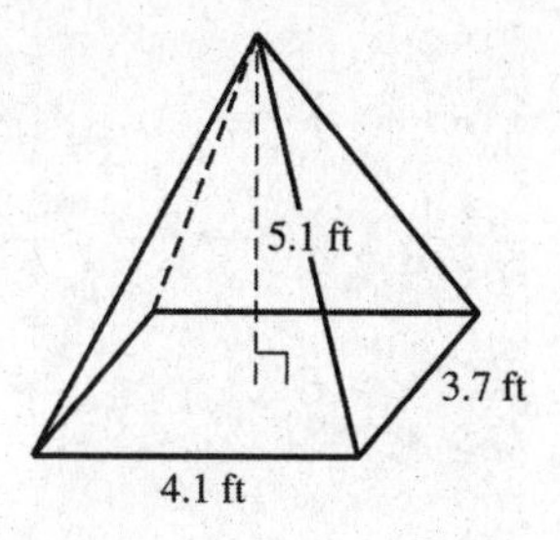

27. ________________

28.

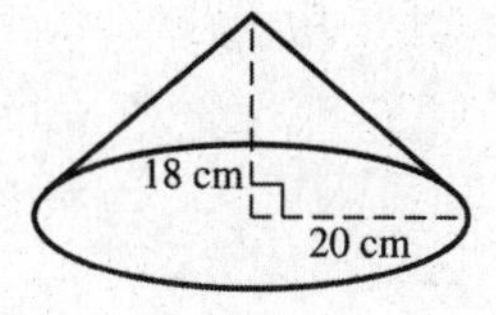

28. ________________

29. Find the volume of a pyramid with square base 42 meters on a side and height 38 meters.

29. ________________

30. Find the volume of a cone with base diameter 3.2 centimeters and height 5.8 centimeters.

30. ________________

Name: Date:
Instructor: Section:

Chapter 8 GEOMETRY

8.8 Pythagorean Theorem

Learning Objectives	
1	Find square roots using the square root key on a calculator.
2	Find the unknown length in a right triangle.
3	Solve application problems involving right triangles.

Key Terms

Use the vocabulary terms listed below to complete each statement in exercises 1–3.

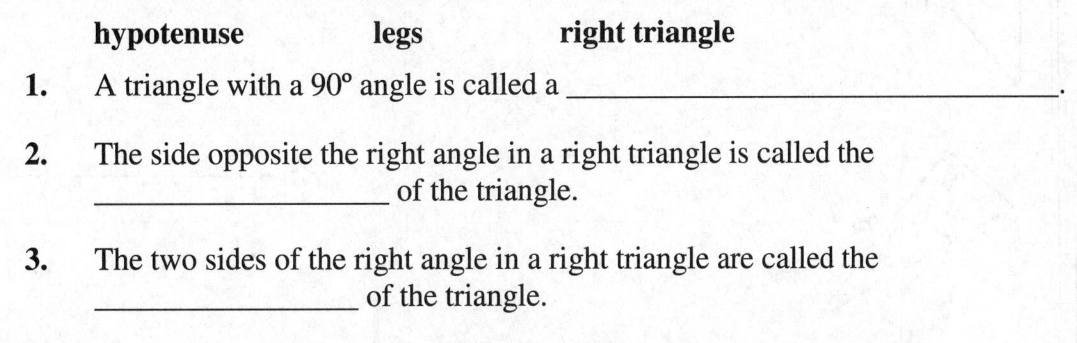

hypotenuse **legs** **right triangle**

1. A triangle with a 90º angle is called a ______________________________.

2. The side opposite the right angle in a right triangle is called the ____________________ of the triangle.

3. The two sides of the right angle in a right triangle are called the __________________ of the triangle.

Objective 1 Find square roots using the square root key on a calculator.

Find each square root. Use a calculator with a square root key. Round the answer to the nearest thousandth, if necessary.

1. $\sqrt{17}$ 1. ________________

2. $\sqrt{27}$ 2. ________________

3. $\sqrt{2}$ 3. ________________

4. $\sqrt{55}$ 4. ________________

5. $\sqrt{75}$ 5. ________________

6. $\sqrt{102}$ 6. ________________

Name: Date:
Instructor: Section:

7. $\sqrt{145}$ 7. ______

Objective 2 Find the unknown length in a right triangle.

Find the unknown length in each right triangle. Use a calculator with a square root key. Round the answer to the nearest tenth, if necessary.

8. 8. ______

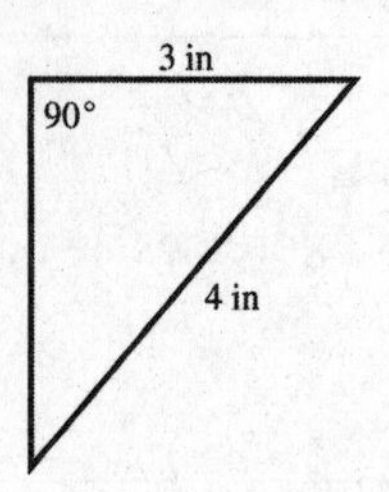

9. 9. ______

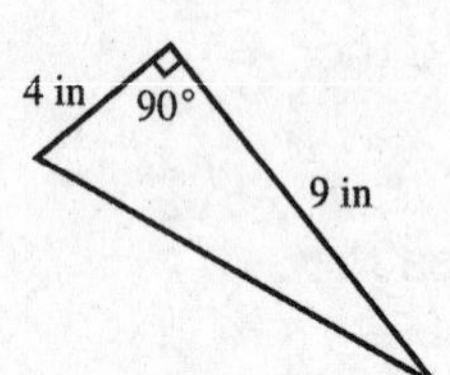

10. 10. ______

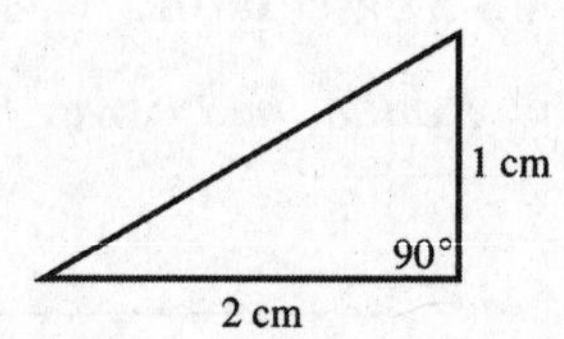

11. 11. ______

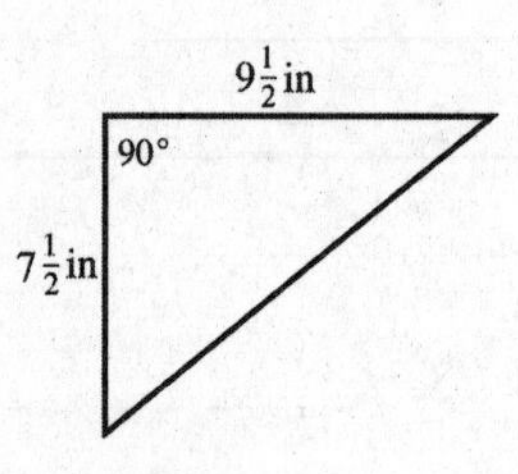

12. 12. ______

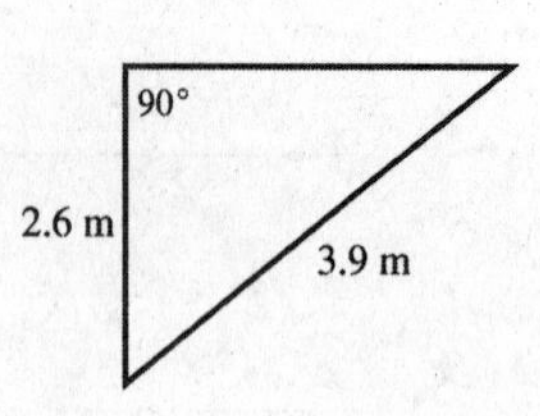

Name: Date:
Instructor: Section:

13.

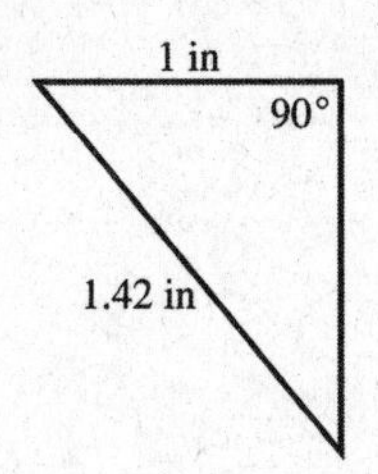

13. ______________

14.

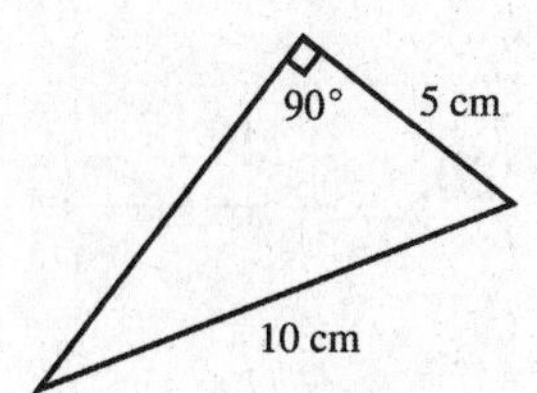

14. ______________

15.

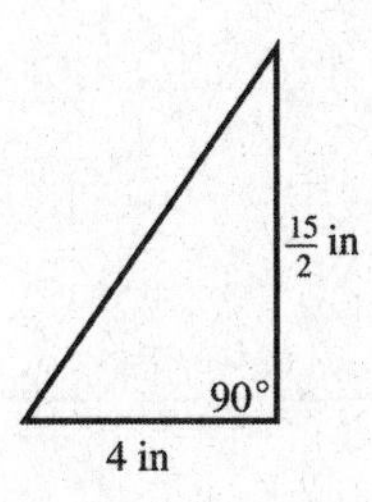

15. ______________

16.

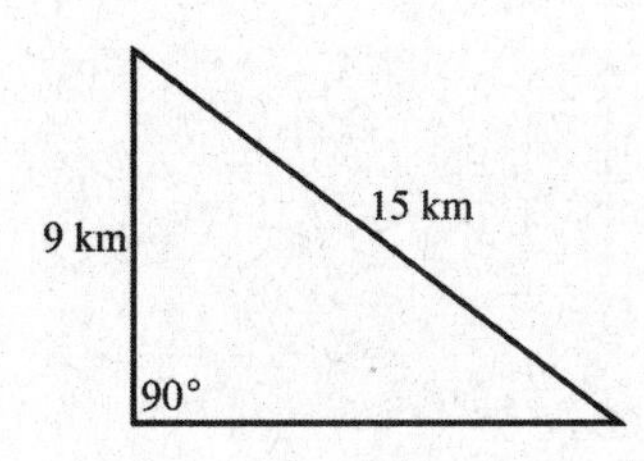

16. ______________

17.

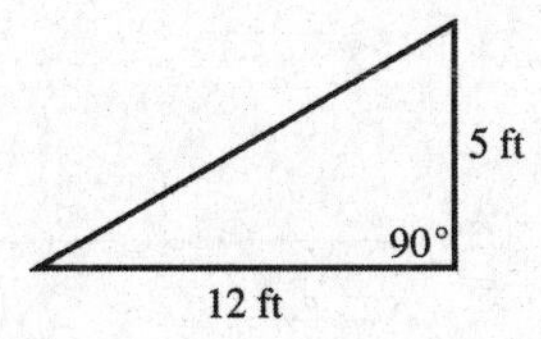

17. ______________

Objective 3 Solve application problems involving right triangles.

Solve each application problem. Draw a diagram if one is not provided. Use a calculator with a square root key. Round the answer to the nearest tenth, if necessary.

18. Find the length of the loading ramp.

18. ______________

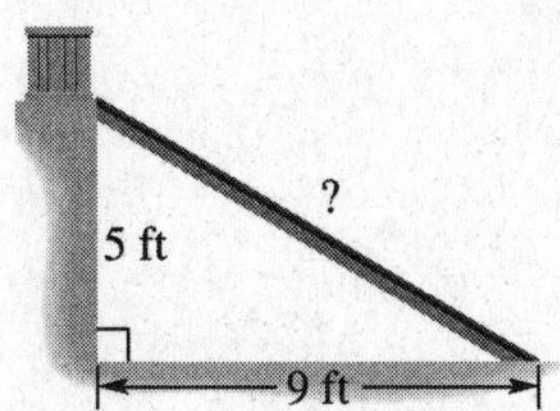

Name: Date:
Instructor: Section:

19. Find the unknown length in this roof plan.

19. 

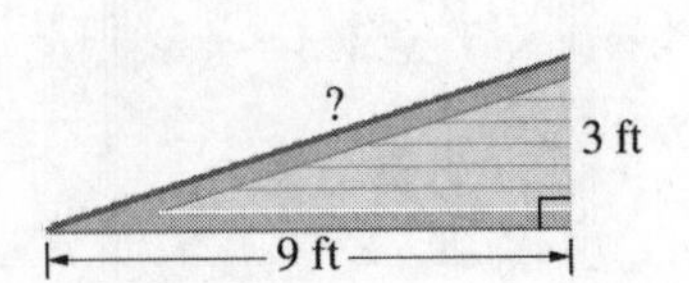

20. A boat travels due south from a dock 10 miles and then turns and travels due east for 15 miles. How far is the boat from the dock?

20. 

21. A boat is pulled into a dock with a rope attached to the bottom of the boat. When the boat is 12 feet from the dock, the length of the rope is 13 feet. How high is the dock?

21. 

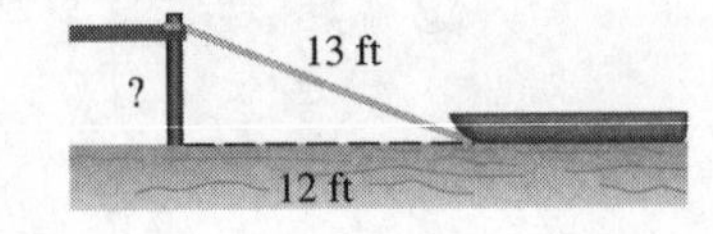

22. A solar panel on a roof is 26 inches wide. It is mounted on a frame whose base is 23.3 inches long. How tall is the frame?

22. 

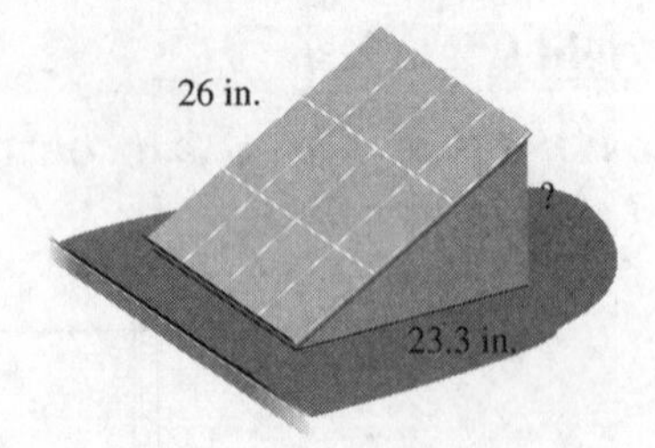

Name: Date:
Instructor: Section:

23. A kite is flying on 50 feet of string. If the horizontal distance of the kite from the person flying it is 40 feet, how far off the ground is the kite?

23. 

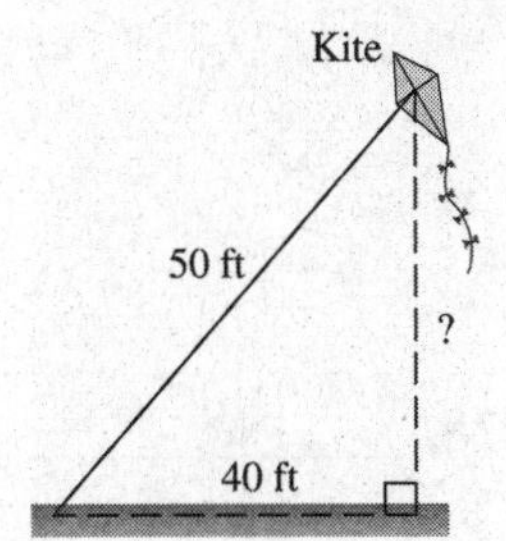

24. Find the distance between the centers of the holes in the metal plate.

24. 

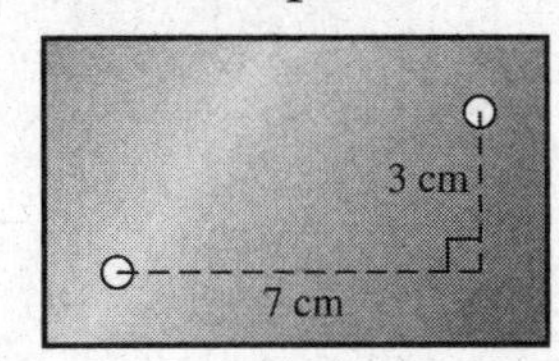

25. The base of a 17-ft ladder is located 15 ft from a building. How high up on the building will the ladder reach?

25. 

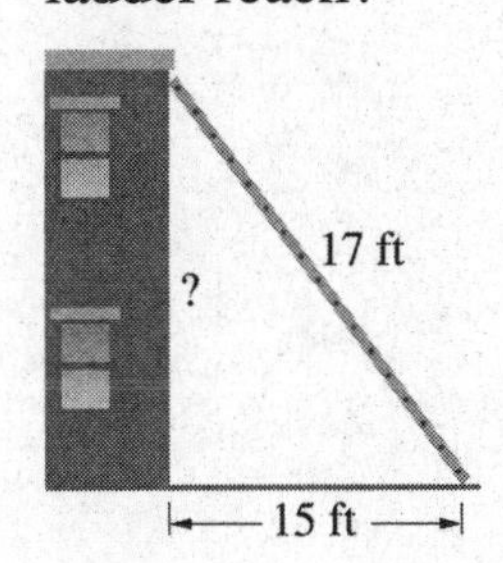

26. The base of a ladder is located 7 feet from a building. The ladder reaches 24 feet up the building. How long is the ladder?

26. 

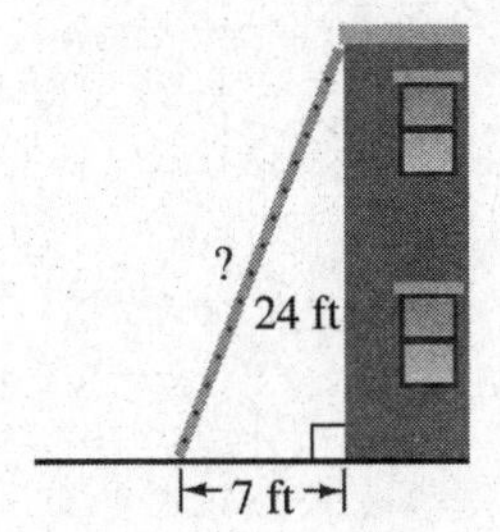

Name: Date:

Instructor: Section:

27. A truck is stopped 6 feet from a door into a storeroom. If the back of the truck is 4 feet above ground level, how long a ramp is needed to unload the truck?

27. 

28. What is the radius of the circle?

28. 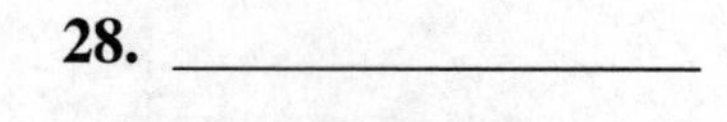

29. A repairman needs to fix the siding located 18 feet up from the ground on a house. Since there are bushes next to the house. The base of the ladder must be 6.5 feet from the house. How long must the ladder be to reach the repair site?

29. ______________

30. An access ramp is being built to a door in a building as shown below. Find the length of the ramp.

30. ______________

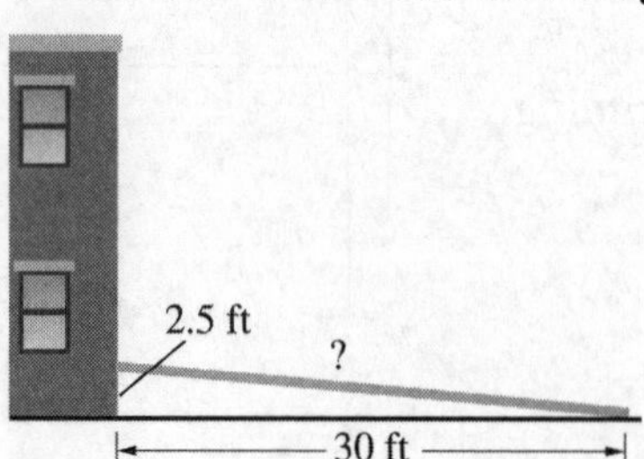

Name: Date:
Instructor: Section:

Chapter 8 GEOMETRY

8.9 Similar Triangles

Learning Objectives
1 Identify corresponding parts in similar triangles.
2 Find the unknown lengths of sides in similar triangles.
3 Solve application problems involving similar triangles.

Key Terms

Use the vocabulary terms listed below to complete each statement in exercises 1–2.

similar triangles **congruent**

1. Two angles are ____________________ if their measures are equal.

2. ____________________________________ are triangles with the same shape but not necessarily the same size.

Objective 1 Identify the corresponding parts in similar triangles.

Name the corresponding angles and the corresponding sides in each pair of similar triangles.

1.

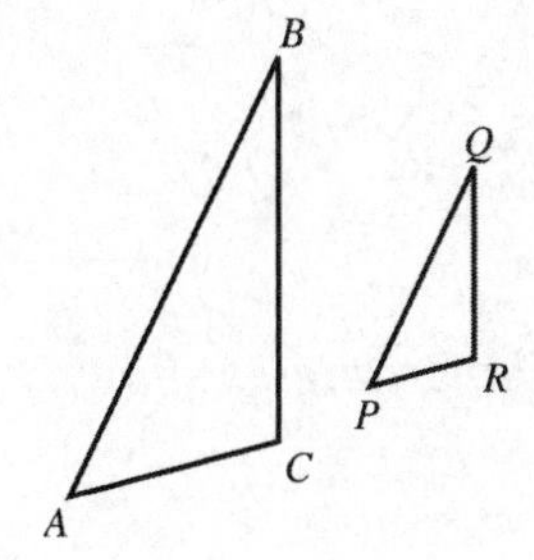

1. ____________________

2.

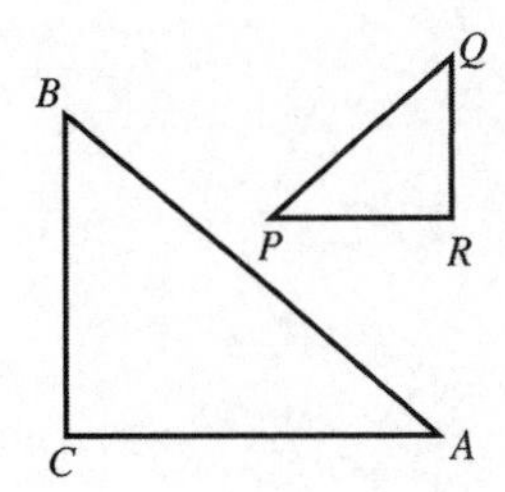

2. ____________________

Name: Date:

Instructor: Section:

3.

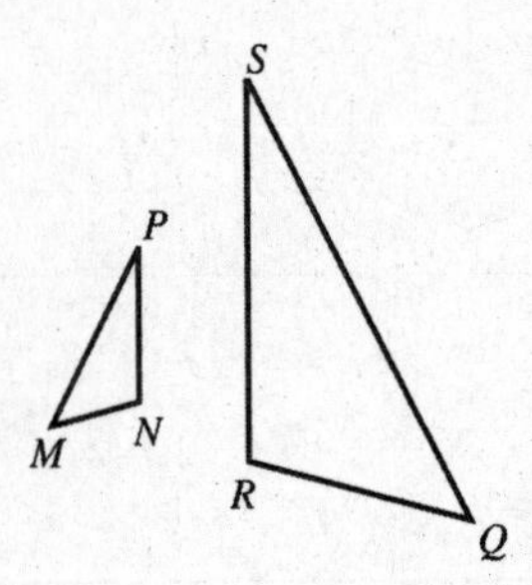

3. ____________

4.

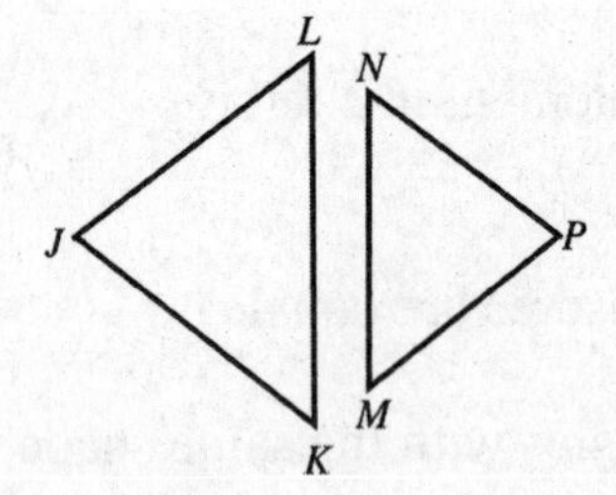

4. ____________

5.

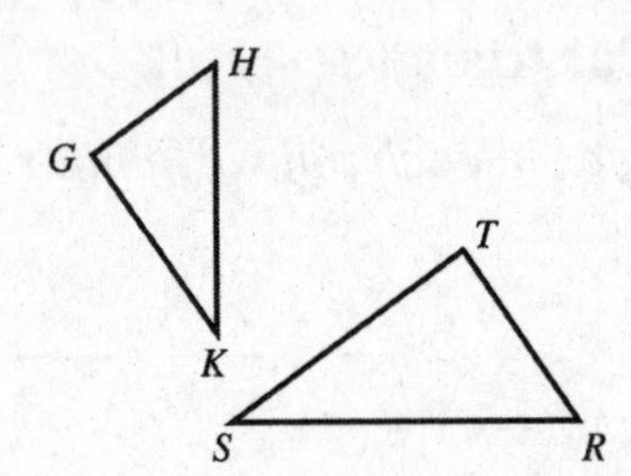

5. ____________

Write the ratio for each pair of corresponding sides in the similar triangles shown below. Write the ratios as fractions in lowest terms.

6. $\dfrac{XY}{NM}, \dfrac{XZ}{NL}, \dfrac{YZ}{ML}$

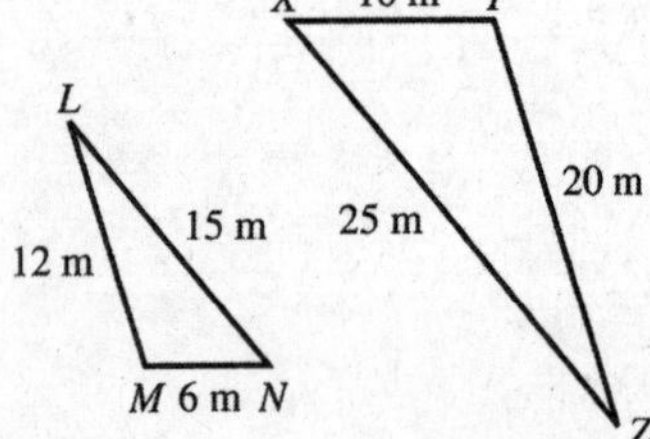

6. ____________

7. $\dfrac{QR}{BA}, \dfrac{SR}{CA}, \dfrac{SQ}{CB}$

7. ____________

B 14 in. C

10 in. 16 in.

A

S

21 in. 24 in.

Q 15 in. R

Name: Date:
Instructor: Section:

8. $\frac{LM}{PS}; \frac{MK}{ST}; \frac{LK}{PT}$

8. ______________

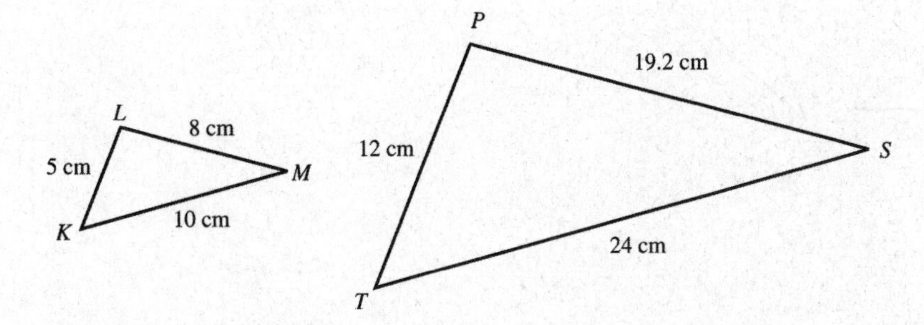

Objective 2 Find the unknown lengths of sides in similar triangles.

Find the unknown lengths in each pair of similar triangles.

9.

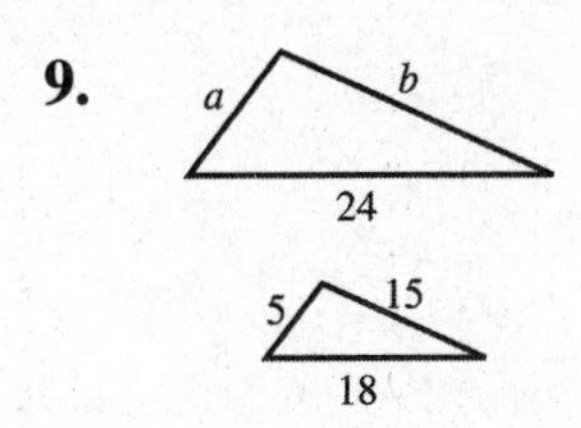

9. ______________

10.

21 21 28
6 a b

10. ______________

11.

18 a b
10 12 6

11. ______________

12.

m 70 50
r 27 15

12. ______________

Name: Date:
Instructor: Section:

13.

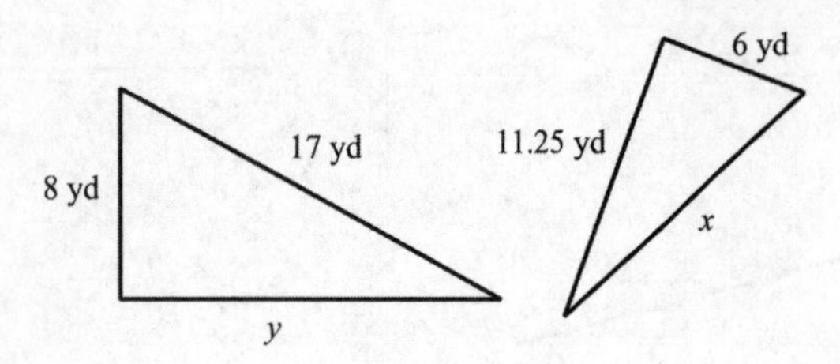

13. ______________

14.

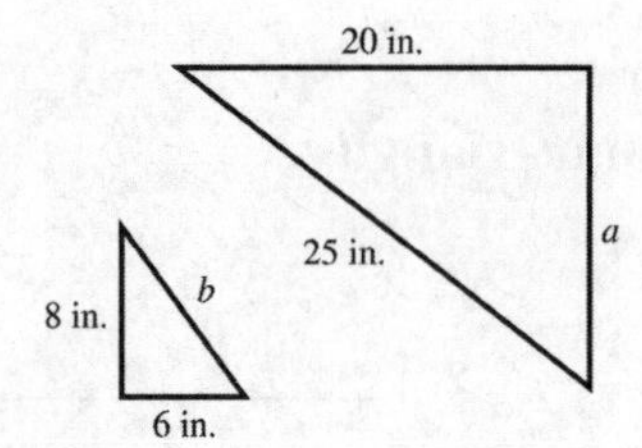

14. ______________

Find the unknown length in each of the following. Round the answer to the nearest tenth, if necessary. Note: When a line is drawn parallel to one side of a triangle, the smaller triangle that is formed will be similar to the original triangle.

15.

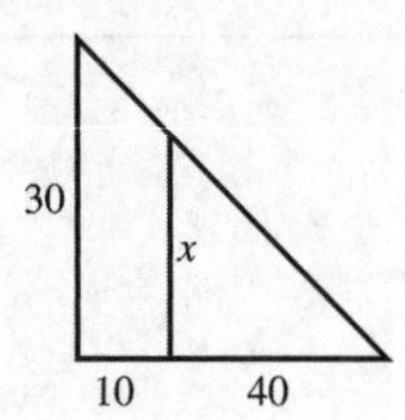

15. ______________

16.

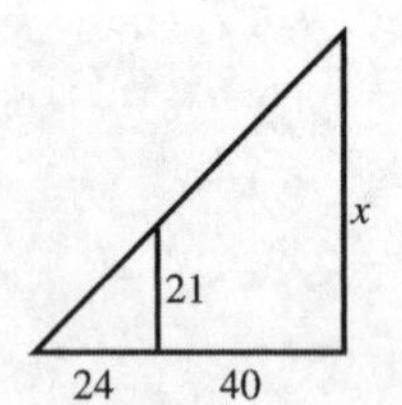

16. ______________

Name: Date:
Instructor: Section:

Find the perimeter of each triangle. Assume the triangles are similar.

17.

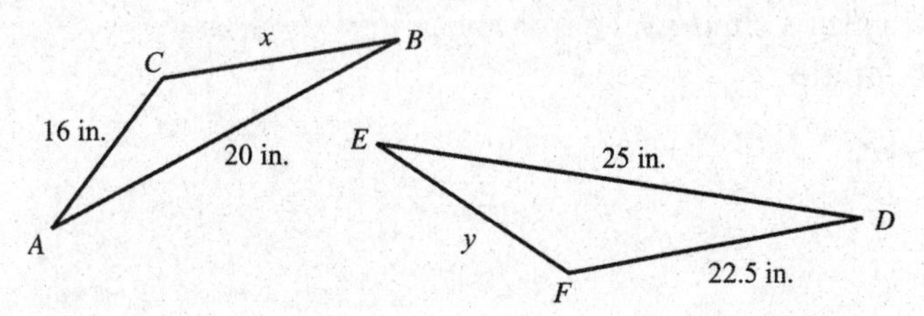

17. *ABC*____________

DEF ____________

18.

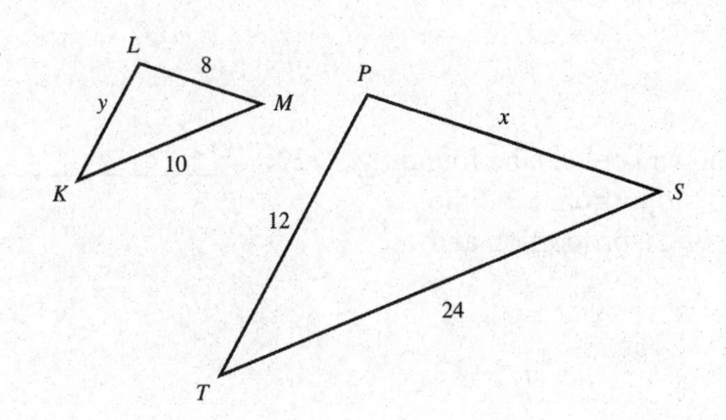

18. *LMK* ____________

PTS ____________

19.

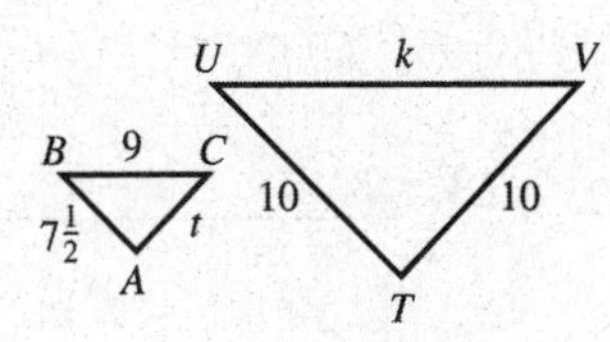

19. *ABC*____________

*TUV*____________

Objective 3 Solve application problems involving similar triangles.

Solve each application problem.

20. A flagpole casts a shadow 77 feet long at the same time that a pole 15 feet tall casts a shadow 55 ft long. Find the height of the flagpole.

20. ____________

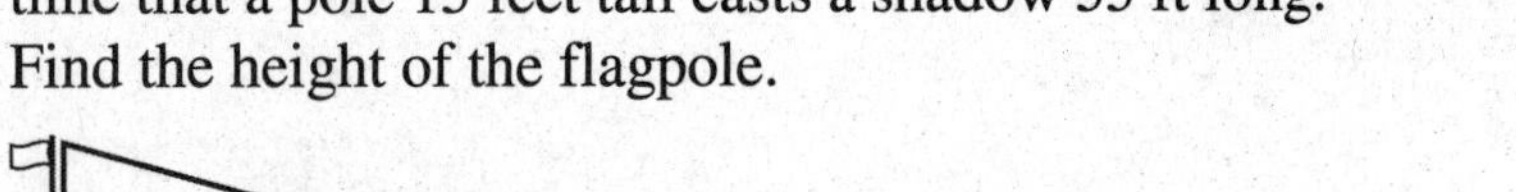

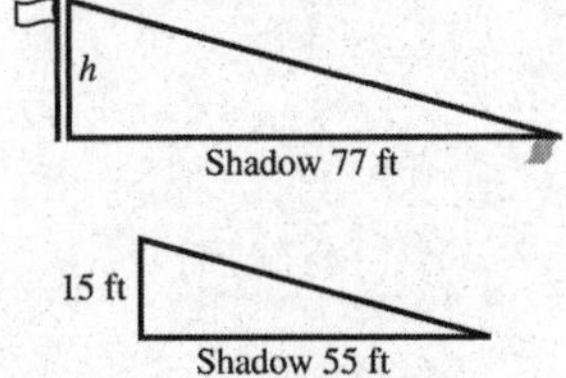

Name: Date:
Instructor: Section:

21. A flagpole casts a shadow 52 m long at the same time that a pole 9 m tall casts a shadow 12 m long. Find the height of the flagpole.

20. 

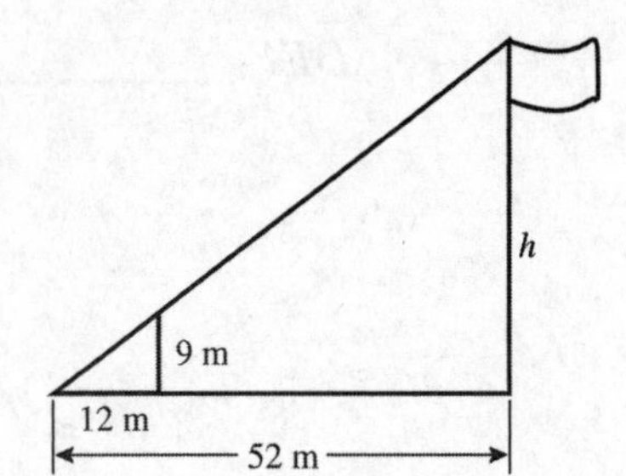

22. The height of the house shown here can be found by using similar triangles and proportion. Find the height of the house by writing a proportion and solving it.

22. 

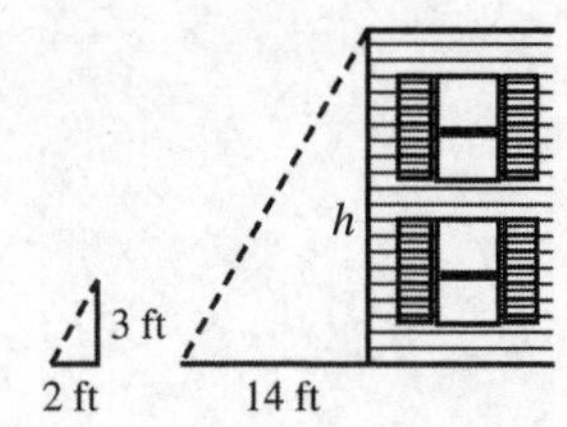

23. A sailor on the USS Ramapo saw one of the highest waves ever recorded. He used the height of the ship's mast, the length of the deck and similar triangles to find the height of the wave. Using the information in the figure, write a proportion and then find the height of the wave.

23. 

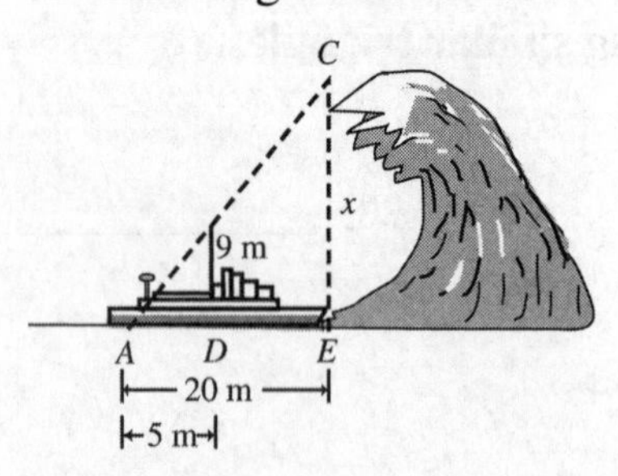

Name: Date:
Instructor: Section:

24. A fire lookout tower provides an excellent view of the surrounding countryside. The height of the tower can be found by lining up the top of the tower with the top of a 3-meter stick. Use similar triangles to find the height of the tower.

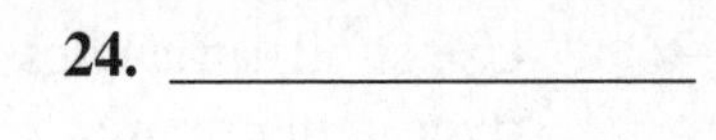

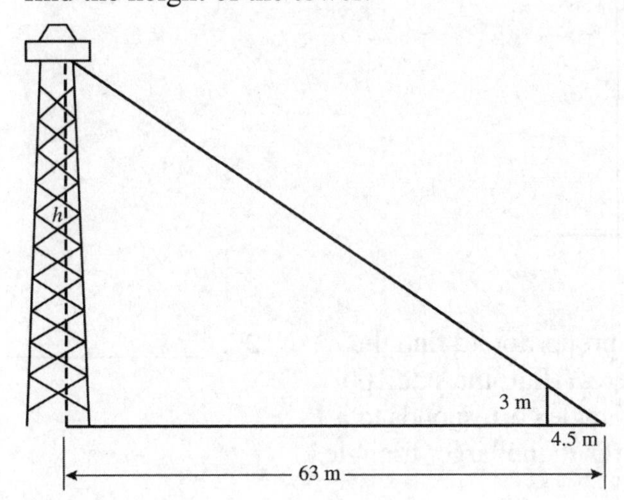

25. The ratio of the rise of a roof to the run of a roof is 5 to 12. Use this information to find the height of the roof indicated by h in the diagram.

25. ______________

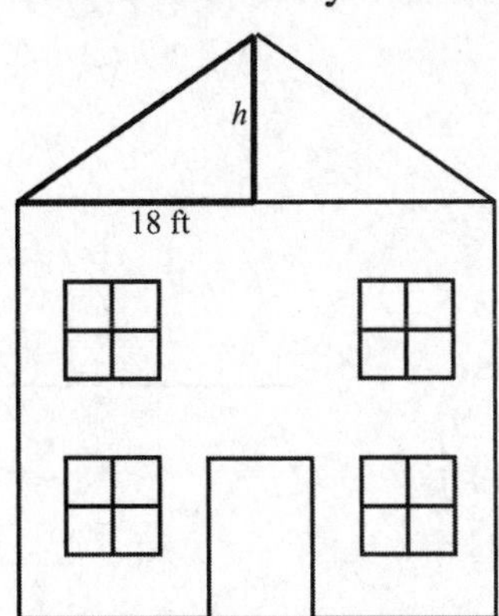

Name: Date:
Instructor: Section:

26. Use similar triangles to find the distance h across the river in the figure.

26. 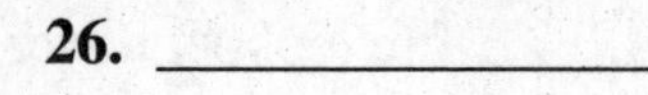

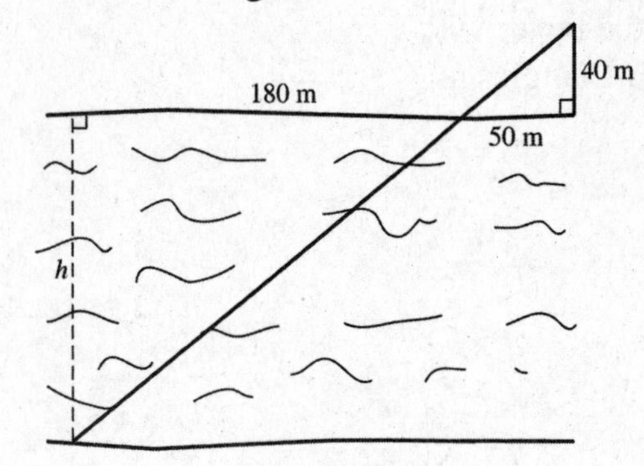

27. Use similar triangles and a proportion to find the length of the lake shown here. (Hint: the side 100 yards long in the smaller triangle corresponds to a side of $100 + 120 = 220$ yards in the larger triangle.)

27. 

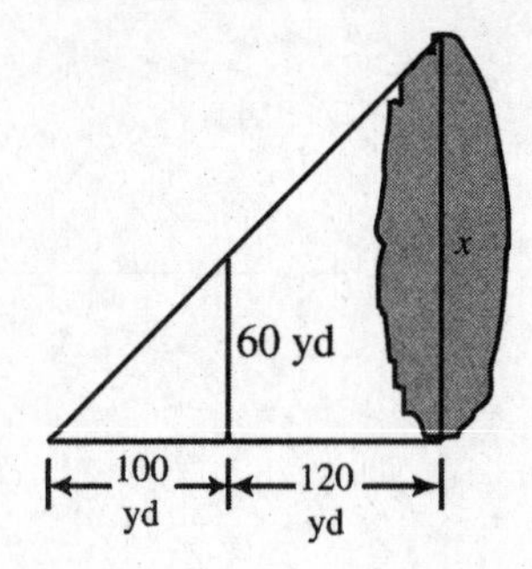

28. Find the height of the tree. (Hint: Since eye-level is 1.8 m above the ground, first find y and then add 1.8 meters for the distance from the ground to eye level.)

28. 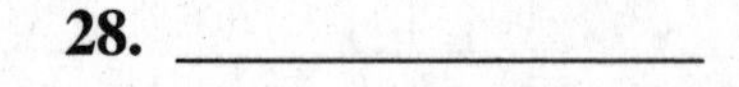

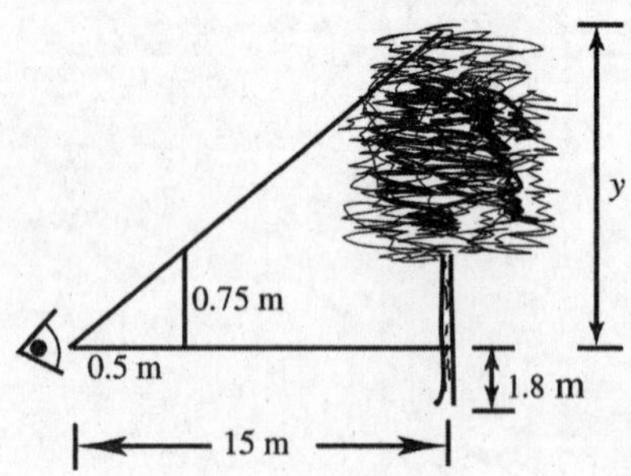

Name: Date:
Instructor: Section:

29. A 30 m ladder touches the side of a building at a height of 25 m. At what height would a 12-m ladder touch the building if it make the same angle with the ground?

29. ______________

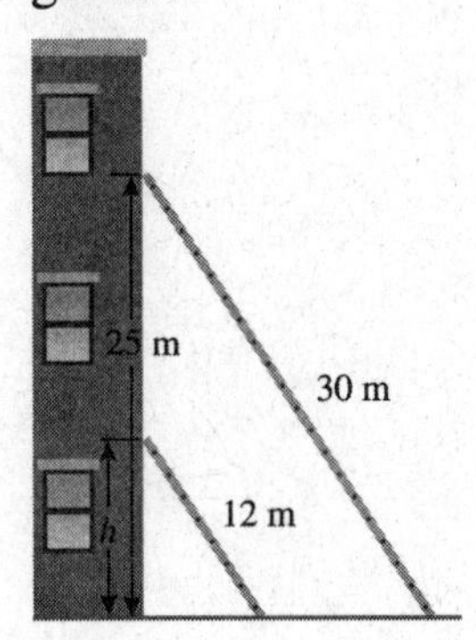

30. Mr. Smith wants to find the height of a tree in his yard. When he holds a yardstick in a vertical position, touching the ground, it casts a shadow 2 ft long. At the same time, the tree's shadow is 32 ft long. What is the height of the tree? (Hint, a yardstick is 3 ft long.)

30. ______________

Name: Date:
Instructor: Section:

Chapter 9 BASIC ALGEBRA

9.1 Signed Numbers

Learning Objectives	
1	Write negative numbers.
2	Graph signed numbers on a number line.
3	Use the > and < symbols.
4	Find absolute value.
5	Find the opposite of a number.

Key Terms

Use the vocabulary terms listed below to complete each statement in exercises 1–4.

negative numbers **signed numbers** **absolute value**

opposite of a number

1. ______________________________ include positive numbers and negative numbers.

2. The ______________________________ is the same distance from 0 on the number line as the original number, but located on the other side of 0.

3. The distance from a number to 0 on the number line is called the ______________________________ of the number.

4. Numbers that are less than 0 are ______________________________.

Objective 1 Write negative numbers.

Write a signed number for each of the following.

1. The temperature was 17 degrees above zero. 1. ________________

2. The stock market fell 120 points. 2. ________________

3. The company had a profit of $830. 3. ________________

4. The temperature was 15 degrees below zero. 4. ________________

5. The lake was 120 feet below sea level. 5. ________________

6. The football team lost 7 yards on the first play. 6. ________________

Name: Date:
Instructor: Section:

Objective 2 Graph signed numbers on a number line.

Graph each set of numbers on a number line.

7. –2, –1, 0, 1, 2, 5

7.

–5 –4 –3 –2 –1 0 1 2 3 4 5

8. –4.5, –1.5, –0.5, 0, 1.5, 2.5

8.

–5 –4 –3 –2 –1 0 1 2 3 4 5

9. $-\frac{1}{2}, -3, -\frac{5}{2}, \frac{1}{4}, 1\frac{7}{8}, 3$

9.

–5 –4 –3 –2 –1 0 1 2 3 4 5

10. $-3, -5, -1\frac{1}{2}, \frac{2}{3}, 0, 4$

10.

–5 –4 –3 –2 –1 0 1 2 3 4 5

11. $-5, -4, -3\frac{2}{3}, -1\frac{1}{2}, 0, 2$

11.

–5 –4 –3 –2 –1 0 1 2 3 4 5

12. $1, 3\frac{7}{9}, 1\frac{1}{4}, -3\frac{1}{8}, -2$

12.

–5 –4 –3 –2 –1 0 1 2 3 4 5

Objective 3 Use the > and < symbols.

Write > or < in each blank to make a true statement.

13. –4____3 **13.** ________________

14. –11____–5 **14.** ________________

15. –23____–32 **15.** ________________

16. 5____–7 **16.** ________________

17. 0____–6 **17.** ________________

18. –5____–3 **18.** ________________

Name: Date:
Instructor: Section:

Objective 4 Find absolute value.

Simplify each absolute value expression.

19. $|-11|$ **19.** ______

20. $-|352|$ **20.** ______

21. $-|-8.23|$ **21.** ______

22. $\left|\frac{8}{9}\right|$ **22.** ______

23. $\left|-\frac{5}{7}\right|$ **23.** ______

24. $-|-7|$ **24.** ______

Objective 5 Find the opposite of a number.

Find the opposite of each number.

25. 3 **25.** ______

26. 8 **26.** ______

27. -3 **27.** ______

28. -10 **28.** ______

29. $-\frac{2}{3}$ **29.** ______

30. $\frac{5}{9}$ **30.** ______

Name: Date:
Instructor: Section:

Chapter 9 BASIC ALGEBRA

9.2 Adding and Subtracting Signed Numbers

Learning Objectives

1	Add signed numbers by using a number line.
2	Add signed numbers without using a number line.
3	Find the additive inverse of a number.
4	Subtract signed numbers.
5	Add or subtract a series of signed numbers.

Key Terms

Use the vocabulary terms listed below to complete each statement in exercises 1–2.

additive inverse **absolute value**

1. $|-4|$ is read as the "______________________________ of negative four."

2. The opposite of a number is its ______________________________.

Objective 1 Add signed numbers by using a number line.

Add by using a number line.

1. $8 + 7$

1. 

−8 −6 −4 −2 0 2 4 6 8

2. $7 + (-12)$

2. ____________________

−8 −6 −4 −2 0 2 4 6 8

3. $-2 + (-6)$

3. ____________________

−8 −6 −4 −2 0 2 4 6 8

4. $-8 + 5$

4. ____________________

−8 −6 −4 −2 0 2 4 6 8

Name: Date:
Instructor: Section:

Objective 2 Add signed numbers without using a number line.

Add.

5. $-10+17$ **5.** ________

6. $-3.2+(-4.7)$ **6.** ________

7. $-7.12+4.3$ **7.** ________

8. $-6.21+7.04$ **8.** ________

9. $-\frac{1}{2}+\frac{5}{4}$ **9.** ________

10. $-\frac{7}{5}+\left(-\frac{7}{10}\right)$ **10.** ________

11. $-4\frac{3}{4}+1\frac{1}{2}$ **11.** ________

12. $3\frac{2}{9}+\left(-2\frac{1}{3}\right)$ **12.** ________

Objective 3 Find the additive inverse of a number.

Give the additive inverse of each number.

13. 0 **13.** ________

14. -15 **14.** ________

15. 281 **15.** ________

16. -4.6 **16.** ________

Name: Date:
Instructor: Section:

17. 2.7

17. ________________

Objective 4 Subtract signed numbers.

Subtract.

18. $4-11$

18. ________________

19. $2.5-3.7$

19. ________________

20. $-1-(-5)$

20. ________________

21. $\frac{2}{3}-\left(-\frac{7}{12}\right)$

21. ________________

22. $-\frac{1}{2}-\left(-\frac{3}{8}\right)$

22. ________________

23. $4.9-(-8.3)$

23. ________________

Objective 5 Add or subtract a series of signed numbers.

Follow the order of operations to work each problem.

24. $-6-(-1)+(-9)$

24. ________________

25. $-3.7-2.5+(-5.3)$

25. ________________

26. $-4.8-(-3.6)+6.4$

26. ________________

Name: Date:
Instructor: Section:

27. $\frac{1}{2}-\frac{2}{3}+\left(-\frac{1}{6}\right)$

27. ________________

28. $\frac{3}{2}+\left(-\frac{1}{3}\right)-\left(-\frac{5}{6}\right)$

28. ________________

29. $5.9-(-4.7)-7.5$

29. ________________

30. $6.8+(-5.9)-(-8.6)$

30. ________________

Name: Date:
Instructor: Section:

Chapter 9 BASIC ALGEBRA

9.3 Multiplying and Dividing Signed Numbers

Learning Objectives
1 Multiply or divide two numbers with opposite signs.
2 Multiply or divide two numbers with the same sign.

Key Terms

Use the vocabulary terms listed below to complete each statement in exercises 1–3.

factors **product** **quotient**

1. Numbers that are being multiplied are called ____________________.
2. The answer to a division problem is called the ____________________.
3. The answer to a multiplication problem is called the ____________________.

Objective 1 Multiply or divide two numbers with opposite signs.

Multiply or divide as indicated.

1. $-6 \cdot 4$ 1. ______________

2. $(-13)(3)$ 2. ______________

3. $8(-12)$ 3. ______________

4. $-\frac{11}{15} \cdot \frac{25}{22}$ 4. ______________

5. $14(-2.3)$ 5. ______________

6. $-7.1(3.8)$ 6. ______________

7. $-6.2(4.8)$ 7. ______________

Name: Date:
Instructor: Section:

8. $\frac{-16}{8}$ **8.** ______________

9. $\frac{28}{-7}$ **9.** ______________

10. $-96 \div 12$ **10.** ______________

11. $-\frac{5}{6} \div \frac{2}{3}$ **11.** ______________

12. $5 \div \left(-\frac{5}{11}\right)$ **12.** ______________

13. $\frac{-59.2}{7.4}$ **13.** ______________

14. $7 \div \left(-\frac{21}{8}\right)$ **14.** ______________

15. $\frac{-11.84}{2}$ **15.** ______________

Objective 2 Multiply or divide two numbers with the same sign.

Multiply or divide as indicated.

16. $-11(-4)$ **16.** ______________

17. $15(6)$ **17.** ______________

Name: Date:
Instructor: Section:

18. $-\frac{1}{4}\cdot(-10)$

18. ______________

19. $-\frac{2}{5}\cdot(-6)$

19. ______________

20. $\frac{2}{5}\cdot(6)$

20. ______________

21. $-15\cdot\left(-\frac{3}{10}\right)$

21. ______________

22. $-\frac{7}{10}\cdot\left(-\frac{5}{4}\right)$

22. ______________

23. $\frac{-\frac{3}{11}}{-\frac{7}{33}}$

23. ______________

24. $-\frac{2}{3}\div -2$

24. ______________

25. $-\frac{7}{18}\div -\frac{7}{9}$

25. ______________

26. $-\frac{5}{8}\div -\frac{10}{3}$

26. ______________

Name: Date:
Instructor: Section:

27. $\dfrac{22.75}{5}$

27. ______________

28. $\dfrac{-14.88}{-3.1}$

28. ______________

29. $-\dfrac{5}{8} \div (-10)$

29. ______________

30. $\dfrac{-\frac{9}{20}}{-\frac{3}{4}}$

30. ______________

Name: Date:
Instructor: Section:

Chapter 9 BASIC ALGEBRA

9.4 Order of Operations

Learning Objectives
1 Use the order of operations.
2 Use the order of operations with exponents.
3 Use the order of operations with fraction bars.

Key Terms

Use the vocabulary terms listed below to complete each statement in exercises 1–3.

exponent **base** **order of operations**

1. For problems or expressions with more than one operation, the ________________________ tells what to do first, second, and so on, to obtain the correct answer.

2. In the expression 3^5, the ________________ is 3.

3. In the expression 3^5, the ________________ is 5.

Objective 1 Use the order of operations.

Simplify.

1. $-5+6+(-3)\cdot 7$ 1. ______________

2. $4+(-3)+2\cdot(-5)$ 2. ______________

3. $6-12\div 3$ 3. ______________

4. $8+3\cdot(-6)$ 4. ______________

5. $(3-9)\cdot(-5)$ 5. ______________

Name: Date:
Instructor: Section:

6. $(11+9)\cdot(-7+3)$ 6. ______

7. $-5\cdot(8-14)\div(-10)$ 7. ______

8. $8\div(-4)+4\cdot(-3)+2$ 8. ______

9. $(-3+5)\cdot(6+1)$ 9. ______

10. $-3\cdot(8-16)\div(-8)$ 10. ______

Objective 2 Use the order of operations with exponents.

Simplify.

11. $36\div(-3)\div 2^2$ 11. ______

12. $-4\cdot 2^2-5\cdot 3-(-6)$ 12. ______

13. $3\cdot 5^2-3\cdot 7-9$ 13. ______

14. $-\left(\frac{3}{7}+\frac{2}{7}\right)+\left(-\frac{1}{3}\right)^2$ 14. ______

15. $2^2\cdot 3^2+(-3)\cdot 2+1$ 15. ______

Name: Date:
Instructor: Section:

16. $\left(-\frac{1}{5}\right)^2-\left(\frac{4}{5}-\frac{1}{5}\right)$

16. ______________

17. $-\left(-\frac{1}{6}+\frac{5}{6}\right)\div\left(-\frac{1}{3}\right)^2$

17. ______________

18. $-\left(-\frac{4}{7}\right)-\left(-\frac{2}{7}\right)\div\left(\frac{1}{7}\right)^2$

18. ______________

19. $6^2\div 3^2-4\cdot 3-2\cdot 5$

19. ______________

20. $-\frac{2}{3}-\frac{1}{3}-\left(\frac{1}{5}\right)^2$

20. ______________

Objective 3 Use the order of operations with fraction bars.

Simplify.

21. $\dfrac{9(-4)}{-6-(-2)}$

21. ______________

22. $\dfrac{4^3-3^3}{-5(-4+2)}$

22. ______________

23. $\dfrac{-4[8-(-3+7)]}{-6[3-(-2)]-3(-3)}$

23. ______________

Name: Date:
Instructor: Section:

24. $\dfrac{5(-8+3)}{13(-2)+(-6-1)(-4+1)}$

24. ________________

25. $\dfrac{3-(-7)-5\cdot 4}{(2-5)^2-(-1)}$

25. ________________

26. $\dfrac{4-(-6)-4\cdot 6}{(3-5)^2-(-3)}$

26. ________________

27. $\dfrac{-9-(-3-4)}{(-3)^2-7}$

27. ________________

28. $\dfrac{-8-(-5-7)}{5-(-3)^2}$

28. ________________

29. $\dfrac{(-4-7)-(-3)^2}{-4-2\cdot 3}$

29. ________________

30. $\dfrac{(-5)^2+(-3-4)}{-3-2\cdot 3}$

30. ________________

Name: Date:
Instructor: Section:

Chapter 9 BASIC ALGEBRA

9.5 Evaluating Expressions and Formulas

Learning Objectives
1 Define variable and expression.
2 Find the value of an expression when values of the variables are given.

Key Terms

Use the vocabulary terms listed below to complete each statement in exercises 1–2.

variable **expression**

1. A combination of operations on variables and numbers is called an ______________________.

2. A letter that represents a number is called a ______________________.

Objective 1 Define variable and expression.

Write if each of the following is a **variable** *or an* **expression**.

1. $3p$ 1. ______________

2. r 2. ______________

3. $7x + 4y$ 3. ______________

Objective 2 Find the value of an expression when values of the variables are given.

Find the value of the expression $3r - 2s$ *for each of the following values of* ***r*** *and s.*

4. $r = 2, s = 5$ 4. ______________

5. $r = -3, s = 4$ 5. ______________

6. $r = 4, s = -6$ 6. ______________

7. $r = -7, s = -3$ 7. ______________

8. $r = 0, s = -12$ 8. ______________

Name: Date:
Instructor: Section:

9. $r = -4, s = 0$ **9.** ____________

Use the given values of the variables to find the value of each expression.

10. $-4k + 3m;\ k = 5,\ m = -\frac{1}{3}$ **10.** ____________

11. $-2d + f;\ d = \frac{1}{2},\ f = -3$ **11.** ____________

12. $\frac{2y - z}{2 - x};\ y = 1,\ z = 6,\ x = 4$ **12.** ____________

13. $\frac{-a + 3b}{c - 2};\ a = 4,\ b = -1,\ c = -5$ **13.** ____________

Evaluate the following expressions if $x = -3,\ y = 2,$ *and* $a = 4$.

14. $-3x + 4y - (a - x)$ **14.** ____________

15. $-x^2 + 3y$ **15.** ____________

16. $2(x - 3)^2 + 2y^2$ **16.** ____________

17. $\frac{2x^2 - 3y}{4a}$ **17.** ____________

18. $\frac{3y^2 + 2x^2}{5x + y^2}$ **18.** ____________

Name: Date:
Instructor: Section:

Using the given values, evaluate each formula. Round to the nearest hundredth, if necessary.

19. $P = a + b + c$; $a = 7, b = 8, c = 5$ **19.**______________

20. $P = 2L + 2W$; $L = 8, W = 6$ **20.**______________

21. $A = \frac{1}{2}bh$; $b = 8, h = 9$ **21.**______________

22. $V = \frac{1}{3}Bh$; $B = 50, h = 3$ **22.** ______________

23. $d = rt$; $r = 65, t = 3$ **23.** ______________

24. $C = 2\pi r$; $\pi \approx 3.14, r = 9$ **24.** ______________

25. $A = \pi r^2$; $\pi = 3.14, r = 6$ **25.** ______________

26. $V = l \times w \times h$; $l = 3\frac{1}{2}, w = 4, h = 1\frac{1}{4}$ **26.** ______________

27. $V = \frac{4}{3}\pi r^3$; $\pi = 3.14, r = 2$ **27.** ______________

28. $s = 2\pi r^2 + 2\pi rh$; $\pi = 3.14, r = 3, h = 5$ **28.** ______________

Name: Date:
Instructor: Section:

29. $C = \frac{5}{9}(F - 32);\ F = 23$

29. ________________

30. $F = \frac{9}{5}C + 32;\ C = 25$

30. ________________

Name: Date:
Instructor: Section:

Chapter 9 BASIC ALGEBRA

9.6 Solving Equations

Learning Objectives

1. Determine whether a number is a solution of an equation.
2. Solve equations using the addition property of equations.
3. Solve equations using the multiplication property of equations.

Key Terms

Use the vocabulary terms listed below to complete each statement in exercises 1–4.

equation **solution** **addition property of equations**

multiplication property of equations

1. An ____________________ is a statement that says two expressions are equal.

2. The ____________________________________ states that both sides of an equation can be multiplied or divided by the same number, except division by 0 is not allowed.

3. The ____________________ of an equation is a number that can replace the variable so that the equation is true.

4. The ____________________________________ states that the same number can be added to or subtracted from both sides of an equation.

Objective 1 Determine whether a number is a solution of an equation.

Decide whether the given number is a solution of the equation.

1. $b-5=18$; 13 — 1. ________________

2. $-3c=-12$; 4 — 2. ________________

3. $-11c=33$; 3 — 3. ________________

4. $2-3m=2$; 0 — 4. ________________

5. $-x=-3$; 3 — 5. ________________

Name: Date:
Instructor: Section:

6. $5+8m=3;\ -1$ **6.** ______________

7. $2z-3=-2;\ \frac{1}{2}$ **7.** ______________

8. $-5y+1=6;\ -1$ **8.** ______________

9. $-6k-3=7;\ \frac{2}{3}$ **9.** ______________

10. $9x+3=0;\ \frac{1}{3}$ **10.** ______________

Objective 2 Solve equations using the addition property of equations.

Solve each equation using the addition property. Check each solution.

11. $p-5=9$ **11.** ______________

12. $y+11=16$ **12.** ______________

13. $-2+m=-1$ **13.** ______________

14. $-12=-10+a$ **14.** ______________

15. $x+\frac{1}{3}=1$ **15.** ______________

Name: Date:
Instructor: Section:

16. $m - \frac{3}{4} = 11$

16. ________________

17. $6 = k + \frac{2}{3}$

17. ________________

18. $s - 4 = \frac{5}{9}$

18. ________________

19. $x - 1.24 = 4.37$

19. ________________

20. $6.99 = a + 3.27$

20. ________________

Objective 3 Solve equations using the multiplication property of equations.

Solve each equation using the multiplication property. Check each solution.

21. $\frac{k}{2} = 16$

21. ________________

22. $\frac{1}{2}p = 6$

22. ________________

23. $-12 = \frac{r}{3}$

23. ________________

24. $-16 = -\frac{4}{5}x$

24. ________________

Name: Date:
Instructor: Section:

25. $1.32 = -1.2m$ **25.** ________________

26. $-3.2y = -8.32$ **26.** ________________

27. $-\frac{1}{4}m = 8$ **27.** ________________

28. $\frac{1}{4} = \frac{2}{3}c$ **28.** ________________

29. $\frac{k}{4.2} = 0.5$ **29.** ________________

30. $-1.1 = \frac{m}{-5.2}$ **30.** ________________

Name: Date:
Instructor: Section:

Chapter 9 BASIC ALGEBRA

9.7 Solving Equations with Several Steps

Learning Objectives
1. Solve equations with several steps.
2. Use the distributive property.
3. Combine like terms.
4. Solve more difficult equations.

Key Terms

Use the vocabulary terms listed below to complete each statement in exercises 1–2.

distributive property **like terms**

1. Terms that have exactly the same variables and exponents are called ______________________.

2. The ________________________________ states that $a(b + c) = ab + bc$.

Objective 1 Solve equations with several steps.

Solve each equation. Check each solution.

1. $5p-4=1$ 1. __________

2. $-2p+5=13$ 2. __________

3. $13=2y-9$ 3. __________

4. $6k+4=16$ 4. __________

5. $0.2x+4.1=3.7$ 5. __________

6. $5p-4.2=-17.7$ 6. __________

Name: Date:
Instructor: Section:

7. $-\frac{1}{2}z+1=-2$ 7. ______________

8. $0.3x-12.5=-14.6$ 8. ______________

Objective 2 Use the distributive property.

Use the distributive property to simplify.

9. $-5(2+a)$ 9. ______________

10. $-3(5-a)$ 10. ______________

11. $\frac{1}{2}(10-2x)$ 11. ______________

12. $-5(7-r)$ 12. ______________

13. $-4(8-x)$ 13. ______________

14. $7(k-5)$ 14. ______________

Objective 3 Combine like terms.

Combine like terms.

15. $-3m+5m$ 15. ______________

16. $9z-4z$ 16. ______________

17. $15a-20a$ 17. ______________

18. $-3.2x + 1.3x$ **18.** ______________

19. $5a - 2.7a$ **19.** ______________

20. $\frac{1}{3}k - \frac{5}{6}k$ **20.** ______________

Objective 4 Solve more difficult equations.

Solve each equation. Check each solution.

21. $6y - 13y = -14$ **21.** ______________

22. $\frac{2}{3}z + 2 = \frac{1}{2}z - 1$ **22.** ______________

23. $26 = 13(2 - a)$ **23.** ______________

24. $\frac{1}{2}(b + 3) = -4$ **24.** ______________

25. $-4.1y - y = -2.3 - 2.8$ **25.** ______________

Name: Date:
Instructor: Section:

26. $-3+5-7=-7a+5a$

26. ________________

27. $-3.7z-0.5=-5.2z+4$

27. ________________

28. $-2.1k+4.7=-1.7k+1.5$

28. ________________

29. $-6(3-2d)=10$

29. ________________

30. $-12=0.2(5-x)$

30. ________________

Name: Date:
Instructor: Section:

Chapter 9 BASIC ALGEBRA

9.8 Using Equations to Solve Application Problems

Learning Objectives
1 Translate word phrases into expressions with variables.
2 Translate sentences into equations.
3 Solve application problems.

Key Terms

Use the vocabulary terms listed below to complete each statement in exercises 1–5.

indicator words **sum** **difference** **product** **quotient**

increased by **less than** **double** **per**

1. Words in a problem that indicate the necessary operations are ________________________.

2. ____________________ and ____________________ are indicator words for addition.

3. ____________________ and ____________________ are indicator words for multiplication.

4. ____________________ and ____________________ are indicator words for division.

5. ____________________ and ____________________ are indicator words for subtraction.

Objective 1 Translate word phrases into expressions with variables.

Write each word phrase in symbols, using ***x*** *as the variable.*

1. The sum of 9 and a number 1. ________________

2. 5 subtracted from a number 2. ________________

3. The product of a number and 3 3. ________________

4. Twice a number added to 7 4. ________________

Name: Date:
Instructor: Section:

5. The sum of three times a number and 3 — 5. ____________

6. Double a number added to five times the number — 6. ____________

7. Six times a number subtracted from ten times the number — 7. ____________

Objective 2 Translate sentences into equations.

Translate each sentence into an equation and solve it. Check your solution by going back to the words in the original problem.

8. If –5 times a number is added to 4, the result is –11. Find the number.

8. **Equation** ____________

Solution ____________

9. If twice a number is subtracted from 45, the result is 35. Find the number.

9. **Equation** ____________

Solution ____________

10. The sum of 6 and four times a number is 50. Find the number.

10. **Equation** ____________

Solution ____________

11. When twice a number is decreased by 3, the result is –17. Find the number.

11. **Equation** ____________

Solution ____________

Name: Date:
Instructor: Section:

12. If the product of some number and 2 is increased by 18, the result is four times the number. Find the number.

12.
Equation ____________

Solution ____________

13. If seven times a number is subtracted from nine times a number, the result is 16. Find the number.

13.
Equation ____________

Solution ____________

14. When two times a number is subtracted from 8, the result is 20 plus the number. Find the number.

14.
Equation ____________

Solution ____________

15. If half a number is added to 4, the result is 10. Find the number.

15.
Equation ____________

Solution ____________

Objective 3 Solve application problems.

Solve each application problem using the six problem-solving steps listed in the text.

16. The price of a DVD is $3.00 less than twice the cost of a book. If the DVD costs $25.00, how much does the book cost?

16. ______________

17. If 2 is subtracted from four times a number, the result is 3 more than nine times the number. What is the number?

17. ______________

Name: Date:
Instructor: Section:

18. The larger of two numbers is twice the smaller number. The sum of the numbers is 36. Find the smaller number.

18. ______________

19. The length of a rectangle is 27 centimeters, while the perimeter is 70 centimeters. Find the width of the rectangle.

19. ______________

20. The length of a rectangle is 3 in. more than the width of the rectangle. If the perimeter of the rectangle is 26 in., find the length and the width of the rectangle.

20. Length__________

Width __________

21. Jerry is three times as old as Marie. The difference of their ages is 24. What are their ages?

21. Jerry __________

Marie__________

22. If half a number is added to twice the number, the result is 55. Find the number.

22. ______________

23. If ten times a number is subtracted from six times the number, the result is 12. Find the number.

23. ______________

Name: Date:
Instructor: Section:

24. When the difference between a number and 4 is multiplied by –3, the result is two more than –5 times the number. Find the number.

24. ______________

25. A board is 91 centimeters long. It is to be cut into two pieces, with one piece 15 centimeters longer than the other. Find the length of the shorter piece.

25. ______________

26. Lisa and Michael were opposing candidates for city council. Lisa won, with 73 more votes than Michael. The total number of votes received by both candidates was 567. Find the number of votes received by Michael.

26. ______________

27. A rental car costs \$32 per day plus \$0.20 per mile. The bill for a one-day rental was \$82. How many miles was the car driven?

27. ______________

28. Mrs. Wong's class read eighteen less than twice as many books as Mr. Lee's class read. If Mrs. Wong's class read 40 books, how many books did Mr. Lee's class read?

28. ______________

Name: Date:
Instructor: Section:

29. A wooden railing is 87 inches long. It is to be divided into four pieces. Three of the pieces will be the same length, and the fourth piece will be 3 inches longer than each of the other three. Find the length that each of the three pieces of equal length will be.

29. ________________

30. How much money must be deposited at 13% per year for 3 years to earn $2730 interest?

30. ________________

Name: Date:
Instructor: Section:

Chapter 10 STATISTICS

10.1 Circle Graphs

Learning Objectives	
1	Read and understand a circle graph.
2	Use a circle graph.
3	Draw a circle graph.

Key Terms

Use the vocabulary terms listed below to complete each statement in exercises 1–2.

circle graph **protractor**

1. A ______________________________ shows how a total amount is divided into parts or sectors.

2. A ______________________________ is a device used to measure the number of degrees in angles or parts of a circle.

Objective 1 Read and understand a circle graph.

The circle graph shows the cost of remodeling a kitchen. Use the graph to answer exercises 1–6.

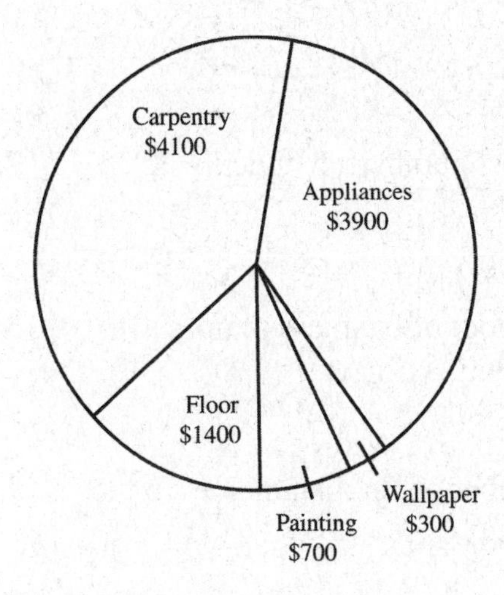

1. Find the total cost of remodeling the kitchen. **1.** ______________

2. What is the largest single expense in remodeling the kitchen? **2.** ______________

3. How much less does the wallpaper cost than painting? **3.** ______________

Name: Date:
Instructor: Section:

4. What fraction of the total cost of remodeling are the appliances? **4.** ________

5. Find the ratio of the cost of wallpaper to the cost of the floor. **5.** ________

6. Find the ratio of the cost of painting to the cost of the floor. **6.** ________

Objective 2 Use a circle graph.

The circle graph shows the number of students enrolled in certain majors at a college. Use the graph to answer exercises 7–12.

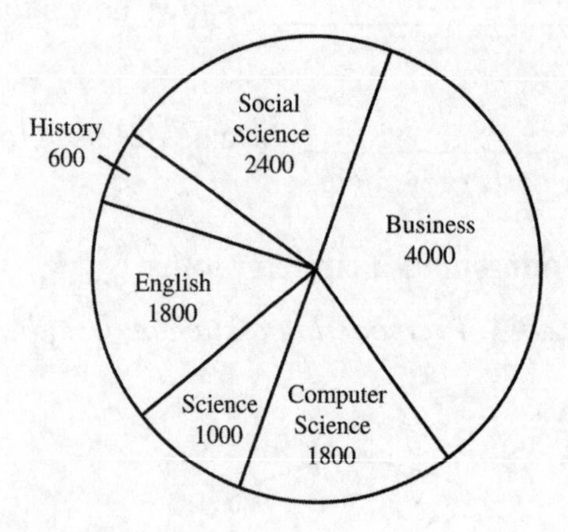

7. Which major has the most number of students enrolled? **7.** ________

8. Find the ratio of the number of business majors to the total number of students. **8.** ________

9. Find the ratio of the number of English majors to the total number of students. **9.** ________

10. Find the ratio of the number of science majors to the number of English majors. **10.** ________

11. Find the ratio of the number of history majors to the number of social science majors. **11.** ________

Name: Date:
Instructor: Section:

12. Find the ratio of the number of computer science majors to the number of business majors.

12. 

The circle graph shows the expenses involved in keeping a sales force on the road. Each expense item is expressed as a percent of the total sales force cost of $950,000. Find the number of dollars of expense for each category in Problems 13–18. Then, use those answers and the circle graph to answer Problems 19–21.

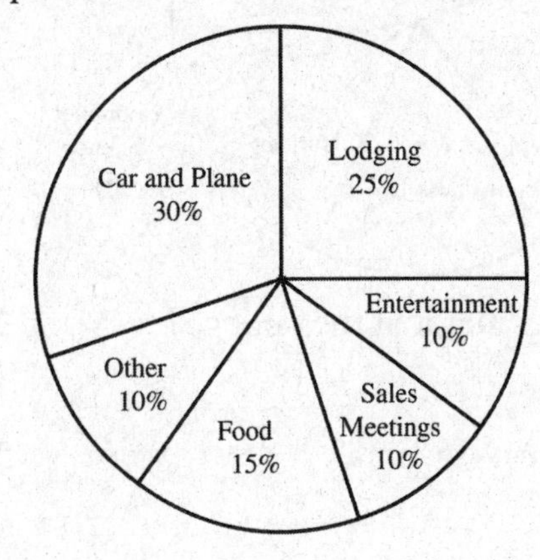

13. Car and plane **13.** ______________

14. Lodging **14.** ______________

15. Entertainment **15.** ______________

16. Sales meetings **16.** ______________

17. Food **17.** ______________

18. Other **18.** ______________

19. What is the ratio of food expense to sales meetings expense? **19.** ______________

20. What percent of the total expenses is spent on food and entertainment? **20.** ______________

21. What is the ratio of car and plane expenses to lodging expenses? **21.** ______________

Name: Date:
Instructor: Section:

The circle graph shows the enrollment by major at a small college. The total enrollment at the college is 3200 students. Use the circle graph to answer questions 22–27.

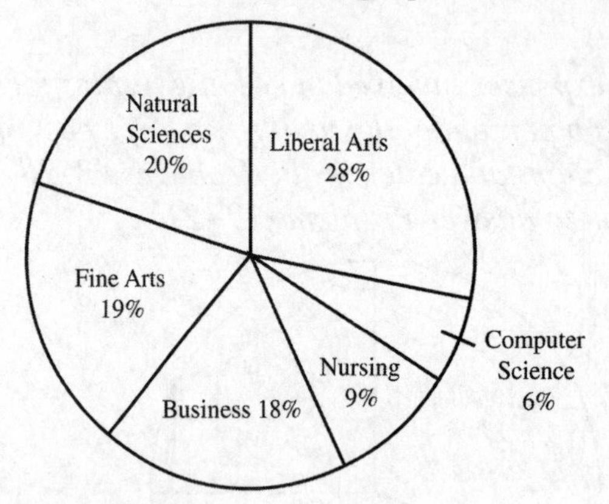

22. What is the most popular major at the college? **22.** ______________

23. What major has the fewest students? **23.** ______________

24. How many business and computer science majors are there in all? **24.** ______________

25. Find the ratio of business majors to computer science majors. **25.** ______________

26. Find the ratio of natural science majors to liberal arts majors. **26.** ______________

27. What percent of the students are either nursing or natural sciences majors? **27.** ______________

Name: Date:
Instructor: Section:

Objective 3 Draw a circle graph.

Use the given information to draw a circle graph.

28. Jensen Manufacturing Company has its annual sales divided into five categories as follows.

Item	*Annual Sales*
Parts	$20,000
Hand tools	80,000
Bench tools	100,000
Brass fittings	140,000
Cabinet hardware	60,000

(a) Find the total sales for a year. **a.** ____________

(b) Find the percent of the total sales for each item.

b. parts ____________

hand tools ________

bench tools _______

brass fittings ______

hardware ________

(c) Find the number of degrees in a circle graph for each item.

c. parts ____________

hand tools ________

bench tools _______

brass fittings ______

hardware ________

Name: Date:
Instructor: Section:

(d) Make a circle graph showing this information.

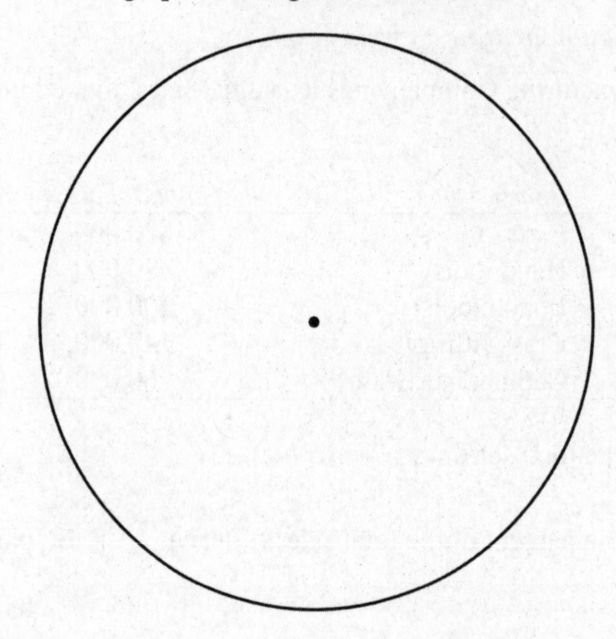

29. A book publisher had 30% of its sales in mysteries, 15% in biographies, 10% in cookbooks, 25% in romance novels, 15% in science, and the rest in business books.

(a) Find the number of degrees in a circle graph for each type of book.

a. mysteries__________

biographies _______

cookbooks________

romance _________

science __________

business__________

(b) Draw a circle graph showing this information.

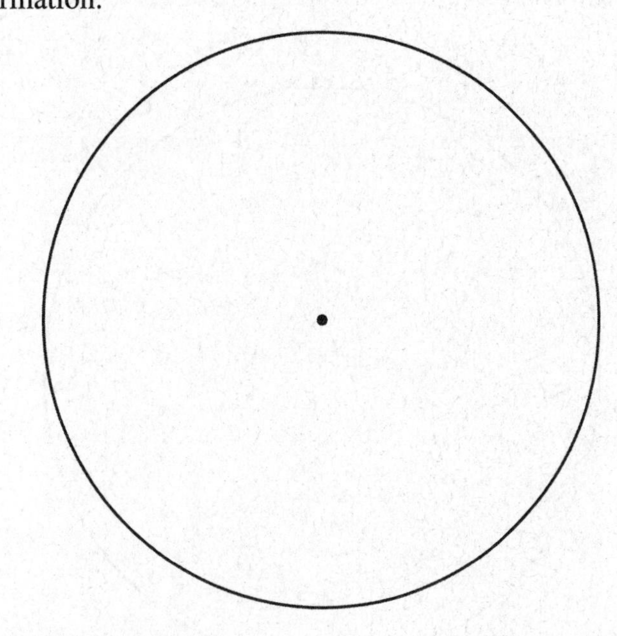

30. A family recorded its expenses for a year, with the following results.

Item	*Percent of Total*
Housing	40%
Food	20%
Automobile	14%
Clothing	8%
Medical	6%
Savings	8%
Other	4%

(a) Find the number of degrees in a circle graph for each item.

a. housing ________

food ________

automobile ________

clothing ________

medical ________

savings ________

other ________

Name: Date:
Instructor: Section:

(b) Draw a circle graph showing this information.

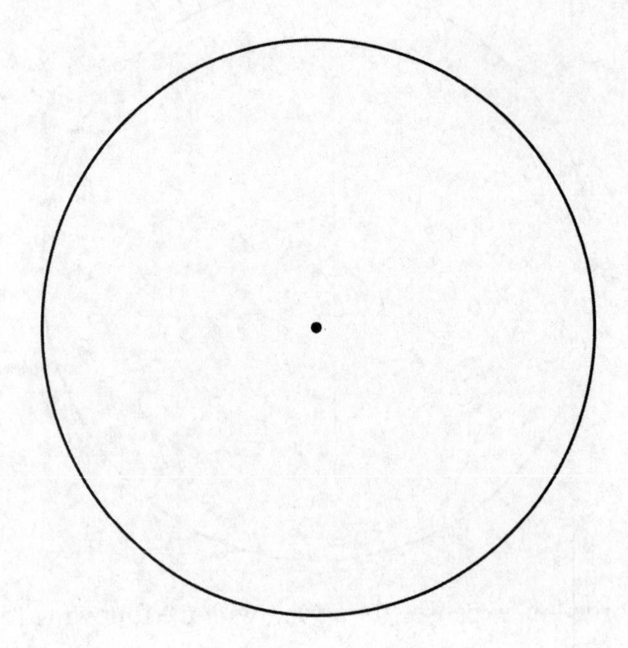

Name: Date:
Instructor: Section:

Chapter 10 STATISTICS

10.2 Bar Graphs and Line Graphs

Learning Objectives
1 Read and understand a bar graph.
2 Read and understand a double-bar graph.
3 Read and understand a line graph.
4 Read and understand a comparison line graph.

Key Terms

Use the vocabulary terms listed below to complete each statement in exercises 1–4.

bar graph **double-bar graph** **line graph** **comparison line graph**

1. A ______________________________ uses dots connected by line to show trends.

2. A ______________________________ compares two sets of data by showing two sets of bars.

3. A ______________________________ uses bars of various heights or lengths to show quantity or frequency.

4. A ______________________________ shows how two sets of data relate to each other by showing a line graph for each item.

Objective 1 Read and understand a bar graph.

The bar graph shows the enrollment for the fall semester at a small college for the past five years. Use this graph for problems 1–7.

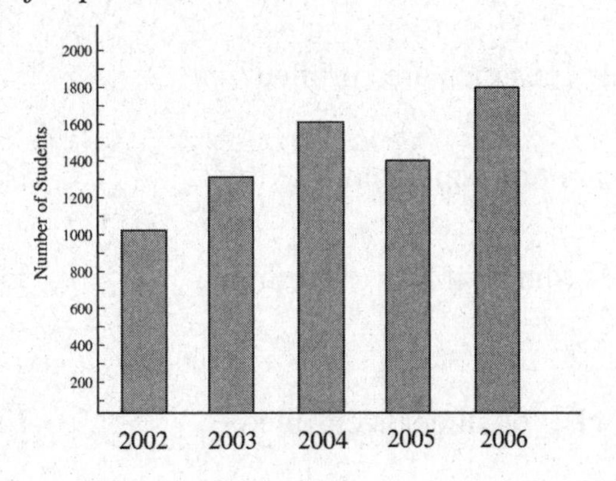

1. What was the fall enrollment for 2002? **1.** ________________

2. What was the fall enrollment for 2004? **2.** ________________

3. What was the fall enrollment for 2006? **3.** ________________

Name: Date:
Instructor: Section:

4. How many more students were enrolled in 2004 than in 2003? **4.** ______________

5. What year had the greatest enrollment? **5.** ______________

6. Which year showed a decrease in enrollment? **6.** ______________

7. By how many students did the enrollment increase from 2005 to 2006? **7.** ______________

Objective 2 Read and understand a double-bar graph.

The double-bar graph shows the enrollment by gender in each class at a small college. Use the double-bar graph for Problems 8–14.

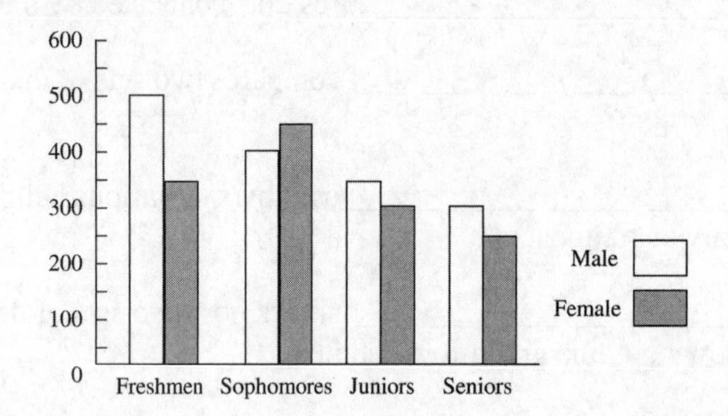

8. Which class has a greater female enrollment than male enrollment? **8.** ______________

9. How many female freshmen are enrolled? **9.** ______________

10. Find the total number of juniors enrolled. **10.** ______________

11. Find the ratio of freshmen males to freshmen females. **11.** ______________

12. Find the total number of students enrolled. **12.** ______________

13. Find the ratio of freshmen students to senior students. **13.** ______________

14. Which class has the greatest difference between male students and female students? **14.** ______________

Name: Date:
Instructor: Section:

Objective 3 Read and understand a line graph.

The line graph gives the value of one share of stock of Microchip Computer Corporation on the first trading day of the month for six consecutive months. Use the line graph for Problems 15–21.

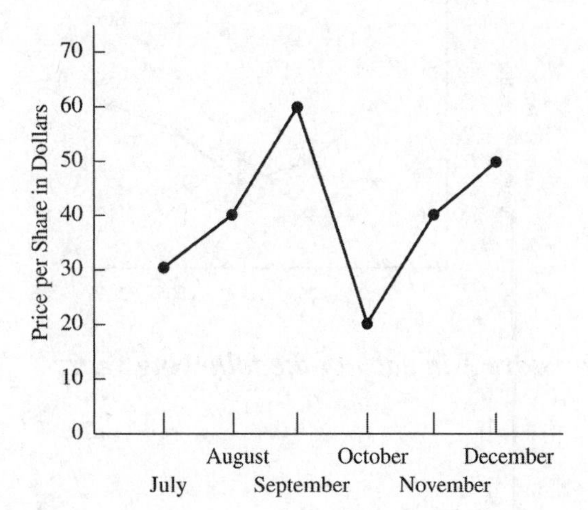

15. In which month was the value of the stock highest? **15.** ______________

16. Find the value of one share on the first trading day October. **16.** ______________

17. Find the increase in the value of one share from October to November. **17.** ______________

18. What is the largest monthly decrease in the value of one share? **18.** ______________

19. Find the ratio of the value of one share on the first trading day in September to the value of one share on the first trading day of October. **19.** ______________

20. Comparing the value of one share on the first trading day in July to the first trading day in November, has the value increased, decreased, or remained unchanged? **20.** ______________

21. By how much did the value of one share increase from July to September? **21.** ______________

Name: Date:
Instructor: Section:

Objective 4 Read and understand a comparison line graph.

The comparison line graph shows annual sales for two different stores for each of the past few years. Use the graph to solve Problems 22–30.

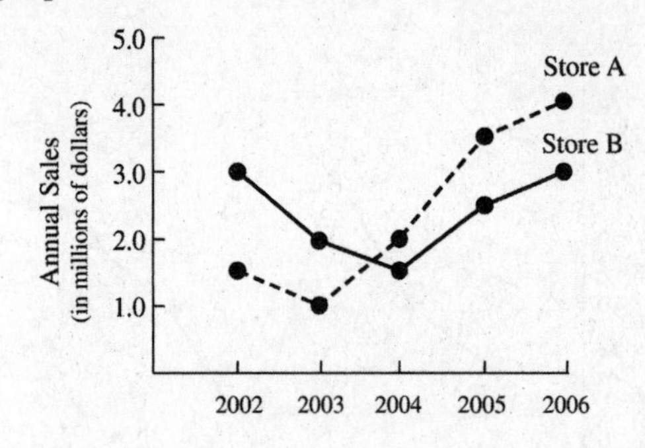

Find the annual sales for store A in each of the following years.

22. 2005 **22.** ____________

23. 2003 **23.** ____________

24. 2002 **24.** ____________

Find the annual sales for store B in each of the following years.

25. 2005 **25.** ____________

26. 2003 **26.** ____________

27. 2002 **27.** ____________

28. In which years did the sales of store *A* exceed the sales of store *B*? **28.** ____________

29. Which year showed the least difference between the sales of *A* and the sales of store *B*? **29.** ____________

30. Find the ratio of the sales of store *A* to the sales of store *B* in 2003. **30.** ____________

Name: Date:
Instructor: Section:

Chapter 10 STATISTICS

10.3 Frequency Distributions and Histograms

Learning Objectives	
1	Understand a frequency distribution.
2	Arrange data in class intervals.
3	Read and understand a histogram.

Key Terms

Use the vocabulary terms listed below to complete each statement in exercises 1–2.

frequency distribution **histogram**

1. A bar graph in which the width of each bar represents a range of number and the height represents the quantity or frequency of items that fall within the interval is called a ________________________.

2. A table that includes a column showing each possible number in the data collected is called a ____________________.

Objective 1 Understand a frequency distribution.

The following scores were earned by students on an algebra exam. Use the data to find the tally and the frequency for the given score in problems 1–6.

84	90	83	72	84	93	83	90	83
90	72	64	90	83	72	83	83	64

1. 64 **1.** tally____________

frequency ________

2. 72 **2.** tally____________

frequency ________

3. 83 **3.** tally____________

frequency ________

4. 84 **4.** tally____________

frequency ________

Name: Date:
Instructor: Section:

5. 90 **5.** tally________

frequency ________

6. 93 **6.** tally________

frequency ________

Objective 2 Arrange data in class intervals.

The following list of numbers represents systolic blood pressure of 21 patients. Use the data to find the tally and the frequency for the given score in problems 7–12. Then answer problems 13 and 14.

120	98	180	128	143	98	105
136	115	190	118	105	180	112
160	110	138	122	98	175	118

7. 90–109 **7.** tally________

frequency ________

8. 110–129 **8.** tally________

frequency ________

9. 130–149 **9.** tally________

frequency ________

10. 150–169 **10.** tally________

frequency ________

11. 179–189 **11.** tally________

frequency ________

12. 190–209 **12.** tally________

frequency ________

13. What was the most common range of systolic blood pressure? **13.** ________________

Name: Date:
Instructor: Section:

14. What was the least common range of systolic blood pressure? **14.** ______________

The following list of numbers represents IQ scores of 18 students. Use these numbers to find the tally and the frequency for the given score in problems 15–19. Then answer problems 20 and 21.

98	121	112	99	105	112
110	100	92	109	104	106
105	88	92	103	98	118

15. 80–89 **15.** tally____________

frequency _______

16. 90–99 **16.** tally____________

frequency _______

17. 100–109 **17.** tally____________

frequency _______

18. 110–119 **18.** tally____________

frequency _______

19. 120–129 **19.** tally____________

frequency _______

20. What was the most common range of IQ scores? **20.** ______________

21. What was the least common range of IQ scores **21.** ______________

Name: Date:
Instructor: Section:

For problems 22 and 23, construct a histogram using the given data.

22. The following list of numbers represents systolic blood pressures of 21 patients. Use intervals 90–109, 110–129, 130–149, 150–169, 170–189, and 190–209.

22. ______________

120	98	180	128	143	98	105
136	115	190	118	105	180	102
160	110	138	122	98	175	118

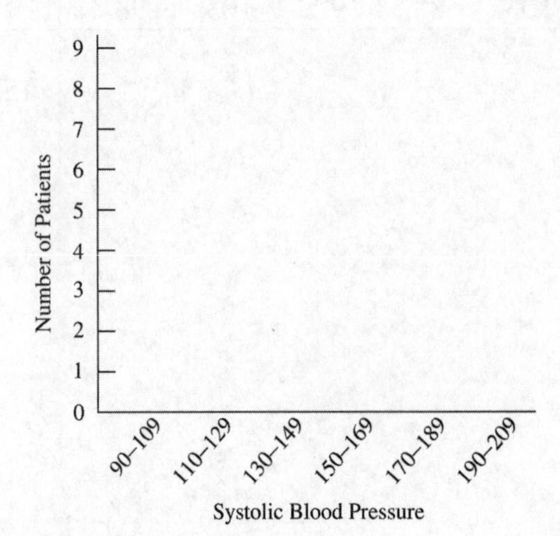

23. The following list of numbers represents IQ scores of 18 students. Use intervals 80–89, 90–99, 100–109, 110–119, and 120–129.

23. ______________

98	121	112	99	105	112
110	100	92	109	104	106
105	88	92	103	98	118

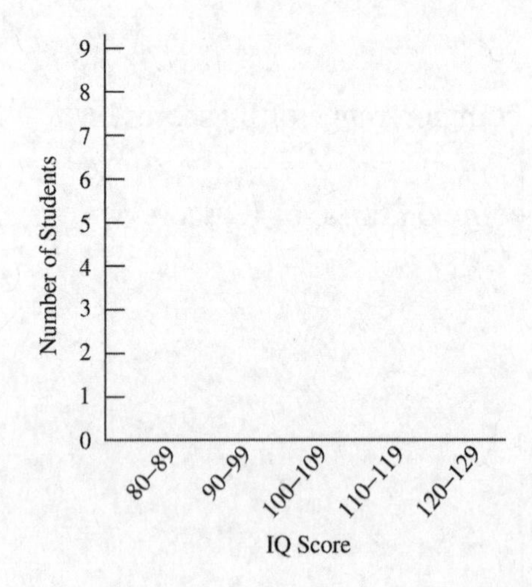

Name: Date:
Instructor: Section:

Objective 3 Read and understand a histogram.

A local chess club recorded the ages of their members and constructed a histogram. Use the histogram to solve Problems 24–30.

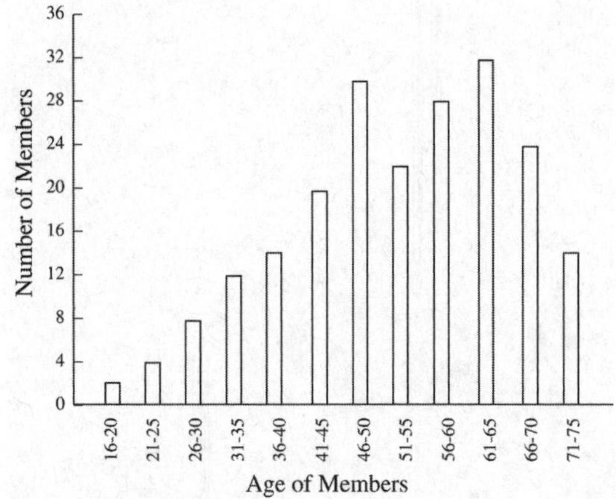

24. The greatest number of members is in which age group? **24.** ______________

25. The fewest number of members are in which age group? **25.** ______________

26. Find the number of members 30 years of age or younger. **26.** ______________

27. Find the number of members 51 years and older. **27.** ______________

28. How many members are 51–65 years of age? **28.** ______________

29. How many members are 46–50 years of age? **29.** ______________

30. Which age range contains the least number of members? **30.** ______________

Name: Date:
Instructor: Section:

Chapter 10 STATISTICS

10.4 Mean, Median, and Mode

Learning Objectives
1 Find the mean of a list of numbers.
2 Find a weighted mean.
3 Find the median.
4 Find the mode.

Key Terms

Use the vocabulary terms listed below to complete each statement in exercises 1–7.

mean **weighted mean** **median** **mode**

bimodal **dispersion** **range**

1. The ______________________________ is the variation or spread of the numbers around the mean.

2. The ______________________________ is the value that occurs most often in a group of values.

3. A mean calculated so that each value is multiplied by its frequency is called a ______________________________.

4. The sum of all the values in a data set divided by the number of values in the data set is called the ______________________________.

5. The middle number in a group of values that are listed from smallest to largest is called the ______________________________.

6. When two values in a data set occur the same number of times, the data set is called ________________________.

7. The difference between the largest value and the smallest value in a set of numbers is called the ____________________________.

Objective 1 Find the mean of a list of numbers.

Find the mean for each list of numbers. Round to the nearest tenth, if necessary.

1. 39, 50, 59, 61, 69, 73, 51, 80 — 1. __________________

2. 31, 37, 44, 51, 52, 74, 69, 83 — 2. __________________

3. 62.7, 59.6, 71.2, 65.8, 63.1 — 3. __________________

Name: Date:
Instructor: Section:

4. 19900, 23850, 25930, 27710, 29340, 41000 **4.** ________

5. 40.1, 32.8, 82.5, 51.2, 88.3, 31.7, 43.7, 51.2 **5.** ________

6. 216, 245, 268, 268, 280, 291, 304, 313 **6.** ________

7. 2.8, 3.9, 4.7, 5.6, 6.5, 9.1 **7.** ________

Objective 2 Find a weighted mean.

Find the weighted mean for each list of numbers. Round to the nearest tenth, if necessary.

8. **8.** ________

Value	*Frequency*
17	4
12	5
15	3
19	1

9. **9.** ________

Value	*Frequency*
13	4
12	2
19	5
15	3
21	1
27	5

10. **10.** ________

Value	*Frequency*
35	1
36	2
39	5
40	4
42	3
43	5

11. **11.** ________

Value	*Frequency*
1	2
2	3
4	5
5	7
6	4
7	2
8	1
9	1

Name: Date:
Instructor: Section:

Find the grade point average for each of the following students. Assume A = 4, B = 3, C = 2, D = 1, F = 0. Round to the nearest tenth, if necessary.

12.

Units	*Grade*
4	C
2	B
5	C
1	D
3	F

12. ______________

13.

Units	*Grade*
3	C
3	A
4	B
5	B
2	A

13. ______________

14.

Units	*Grade*
3	A
4	B
2	C
5	C
2	D

14. ______________

15.

Units	*Grade*
5	B
4	C
3	B
2	C
2	C

15. ______________

Objective 3 Find a median.

Find the median for each list of numbers.

16. 199, 472, 312, 298, 254 **16.** ______________

17. 200, 215, 226, 238, 250, 283 **17.** ______________

18. 0.002, 0.004, 0.012, 0.008 **18.** ______________

19. 389, 464, 521, 610, 654, 672, 682, 712 **19.** ______________

20. 43, 69, 108, 32, 51, 49, 83, 57, 64 **20.** ______________

21. 21, 32, 27, 23, 25, 29, 22 **21.** ______________

Name: Date:
Instructor: Section:

22. 1.8, 1.2, 1.1, 1.9, 2.6 **22.** ________________

23. 200, 195, 302, 284, 256, 237, 239, 240 **23.** ________________

Objective 4 Find the mode.

Find the mode for each list of numbers.

24. 32, 43, 57, 43, 59, 43, 57 **24.** ________________

25. 4, 9, 3, 4, 7, 3, 2, 3, 9 **25.** ________________

26. 238, 272, 274, 272, 268, 271 **26.** ________________

27. 37, 24, 35, 35, 24, 38, 39, 28, 27, 39 **27.** ________________

28. 172.6, 199.7, 182.4, 167.1, 172.6, 183.4, 187.6 **28.** ________________

29. 2, 4, 6, 6, 8, 10, 8, 12, 14, 8 **29.** ________________

30. 0.2, 0.7, 0.9, 0.7, 0.5, 0.3, 0.4, 0.7, 0.2 **30.** ________________

Chapter 1 WHOLE NUMBERS

1.1 Reading and Writing Whole Numbers

Key Terms

1. table
2. whole numbers
3. place value

Objective 1

1. whole number
3. not a whole number

Objective 2

5. 9; 4
7. 1; 1
9. 2; 0
11. 75; 229; 301
13. 300; 459; 200; 5

Objective 3

15. thirty-nine thousand, fifteen
17. two million, fifteen thousand, one hundred two
19. 4127
21. 685,000,259
23. 7210
25. 15,313

Objective 4

27. 177
29. 315

1.2 Adding Whole Numbers

Key Terms

1. commutative property of addition
2. addends
3. addition
4. associative property of addition
5. regrouping (carrying)
6. perimeter
7. sum

Objective 1

1. 12

Objective 2

3. 25
5. 31

Objective 3

7. 99
8. 988
9. 98,977
10. 67,899

Objective 4

11. 112
13. 15,815
15. 4322

Objective 5

17. 38 miles
19. 44 miles
21. 625 tickets
23. 310 feet
25. 1044 yards
27. incorrect; 17,280
29. correct

1.3 Subtracting Whole Numbers

Key Terms

1. minuend
2. regrouping (borrowing)
3. subtrahend
4. difference

Objective 1

1. 187 – 38 = 149; 187 – 149 = 38

3. 785 + 426 = 1211

Objective 2

5. minuend: 98; subtrahend: 38; difference: 62

Objective 3

7. 5151

Objective 4

9. not correct; 153
11. not correct; 2980
13. not correct; 78,087

Objective 5

15. 192
17. 25,899
19. 245

Objective 6

21. 25 boxes
23. $419
25. $5184
27. 2758 people
29. 76 athletes

1.4 Multiplying Whole Numbers

Key Terms

1. commutative property of multiplication
2. factors
3. associative property of multiplication
4. multiple
5. product

Objective 1

1. factors: 5, 2; product: 10

Objective 2

3. 32
5. 192

Objective 3

7. 815
9. 285,867

Objective 4

11. 439,000
13. 6,010,000
15. 16,468,000

Objective 5

17. 193,488
19. 330,687

Objective 5

21. 560 yards
23. $912
25. 576 miles
27. $9728
29. $1602

1.5 Dividing Whole Numbers

Key Terms

1. remainder 2. dividend 3. quotient

4. divisor 5. short division

Objective 1

1. $3\overline{)15}$ with quotient 5; $\dfrac{15}{3} = 5$

Objective 2

3. dividend: 63; divisor: 7; quotient: 9

5. dividend: 44; divisor: 11; quotient: 4

Objective 3, Objective 4

7. undefined 9. undefined 1. undefined

Objective 5, Objective 6

13. 38

Objective 7

15. 144 R 4 17. 141 R 7 19. 170 R 2

Objective 8

21. correct 23. incorrect; 5814 25. correct

Objective 9

27. 2: no; 3: yes; 5: no; 10: no

29. 2: no; 3: no; 5: yes; 10: no

1.6 Long Division

Key Terms

1. dividend
2. remainder
3. divisor
4. quotient
5. long division

Objective 1

1. 82
3. 77
5. 102
7. 85 R84
9. 309 R1
11. 6354 R7
13. 32,587
15. 654 R22

Objective 2

17. 7
19. 6
21. 42
23. 800

Objective 3

25. correct
27. 45 R23
29. incorrect; 296 R79

1.7 Rounding Whole Numbers

Key Terms

1. front end rounding
2. rounding
3. estimate

Objective 1

1. $257{,}3\underline{0}1$
3. $6\underline{4}5{,}371$

Objective 2

5. 7900
7. 18,200
9. 8400
11. 52,000
13. 53,600
15. 600,000

Objective 3

17. $40 + 20 + 60 + 90 = 210$; 210
19. $70 - 40 = 30$; 27
21. $300 + 300 + 200 + 900 = 1700$; 1698
23. $1000 - 400 = 600$; 589
25. $900 \times 800 = 720{,}000$; 715,008
27. $600 + 40 + 200 + 2000 = 2840$; 3280
29. $1000 \times 40 = 40{,}000$; 36,260

1.8 Exponents, Roots, and Order of Operations

Key Terms

1. order of operations
2. square root
3. perfect square

Objective 1

1. exponent: 2; base: 7; 49
3. exponent: 3; base: 8; 512

Objective 2

5. 4
7. 11
9. 15
11. 2500; 2500
13. 400; 20

Objective 3

15. 39
17. undefined
19. 27
21. 54
23. 45
25. 75
27. 28
29. 45

1.9 Reading Pictographs, Bar Graphs, and Line Graphs

Key Terms

1. line graph
2. pictograph
3. bar graph

Objective 1

1. Georgia
3. Minnesota
5. 1%
7. 3
9. Spanish

Objective 2

11. 10
13. 4
15. 18
17. 2900
19. freshman

Objective 3

21. The net sales are increasing every year.
23. 2007
25. 2003–2004
27. 2005
29. $1.5 million

1.10 Solving Application Problems

Key Terms

1. indicator words
2. sum; increased by
3. product; times
4. quotient; per
5. difference; fewer

Objective 1

1. subtraction
3. multiplication
5. subtraction

Objective 2

7. 4 strips
9. 1011 vehicles
11. 6850 people
13. 589 deer
15. 80,128,450 gallons
17. \$3826
19. \$1854

Objective 3

21. $600 \div 60 = 10$ hours; 11 hours

23. $40 \times 30 = 1200$ miles; 936 miles

25. $40{,}000 \times 10 = 400{,}000$ square feet; 217,800 square feet

27. $(20 \times 10) \div 10 \times \$20 = \$400$; \$552

29. $300 - (40 + 20 + 80) + (20 + 40 + 100) = 320$ machines; 348 machines

Chapter 2 MULTIPLYING AND DIVIDING FRACTIONS

2.1 Basics of Fractions

Key Terms

1. improper fraction
2. numerator
3. proper fraction
4. denominator

Objective 1 In each answer, the first fraction is the shaded portion, and the second fraction is the unshaded portion.

1. $\frac{3}{8}; \frac{5}{8}$
3. $\frac{5}{6}; \frac{1}{6}$
5. $\frac{5}{8}; \frac{3}{8}$
7. $\frac{8}{5}; \frac{2}{5}$
9. $\frac{7}{10}; \frac{3}{10}$

Objective 2

11. N: 4; D: 3
13. N: 8; D:11
15. N: 19; D:50
17. N: 19; D: 8
19. N: 157; D: 12

Objective 3

21. improper
23. proper
25. proper
27. improper
29. proper

2.2 Mixed Numbers

Key Terms

1. proper fraction
2. mixed number
3. improper fraction
4. whole numbers

Objective 1

1. $2\frac{1}{2}, 1\frac{1}{6}$

3. none

5. $4\frac{3}{4}$

Objective 2

7. $\frac{11}{6}$

9. $\frac{39}{7}$

11. $\frac{25}{4}$

13. $\frac{15}{2}$

15. $\frac{38}{7}$

17. $\frac{79}{9}$

Objective 3

19. $1\frac{5}{9}$

21. $3\frac{2}{9}$

23. $4\frac{1}{5}$

25. $2\frac{7}{9}$

27. $30\frac{2}{3}$

29. $44\frac{1}{17}$

2.3 Factors

Key Terms

1. factorizations
2. composite number
3. prime factorization
4. prime number
5. factors

Objective 1

1. 1, 7
3. 1, 7, 49
5. 1, 2, 5, 10
7. 1, 5, 25
9. 1, 2, 3, 5, 6, 10, 15, 30

Objective 2

11. neither
13. composite
15. composite
17. prime
19. prime

Objective 3

21. $2^2 \cdot 3$
23. $2^2 \cdot 7$
25. $2^3 \cdot 3$
27. $2^2 \cdot 3^3$
29. $2 \cdot 3^2 \cdot 5^2$

2.4 Writing a Fraction in Lowest Terms

Key Terms

1. lowest terms
2. common factor
3. equivalent fractions

Objective 1

1. no
3. no
5. yes
7. $\frac{1}{3}$
9. $\frac{2}{9}$
11. $\frac{2}{7}$
13. $\frac{3}{22}$

Objective 2

15. $\frac{3 \cdot 3 \cdot 7}{2 \cdot 2 \cdot 3 \cdot 7} = \frac{3}{4}$
17. $\frac{2 \cdot 2 \cdot 3 \cdot 3 \cdot 5}{2 \cdot 3 \cdot 5 \cdot 7} = \frac{6}{7}$
19. $\frac{2 \cdot 2 \cdot 3 \cdot 3}{2 \cdot 3 \cdot 3 \cdot 3} = \frac{2}{3}$
21. $\frac{3 \cdot 5 \cdot 5}{2 \cdot 2 \cdot 5 \cdot 5 \cdot 5} = \frac{3}{20}$

Objective 3

23. equivalent
25. not equivalent
27. equivalent
29. not equivalent

2.5 Dividing Whole Numbers

Key Terms

1. common factor
2. denominator
3. multiplication shortcut
4. numerator

Objective 1

1. $\frac{35}{54}$ 3. $\frac{55}{24}$ 5. $\frac{5}{12}$ 7. $\frac{1}{96}$

Objective 2

9. $\frac{5}{12}$ 11. $\frac{2}{3}$ 13. $\frac{1}{6}$ 15. $\frac{1}{7}$

Objective 3

17. $\frac{4}{5}$ 19. 42 21. 9 23. $1\frac{1}{2}$

Objective 4

25. $\frac{3}{8}\ \text{m}^2$ 27. $\frac{15}{32}\ \text{m}^2$ 29. $\frac{5}{9}\ \text{yd}^2$

2.6 Applications of Multiplication

Key Terms

1. indicator words
2. product

Objective 1

1.	1680 paperbacks	3.	$1500	5.	133 employees
7.	$500	9.	$104	11.	320 muffins
13.	160 pages	15.	90 games	17.	1688 votes
19.	312 square feet	21.	6 gallons	23.	$18,000
25.	$3375	27.	$16,000	29.	$36,000

2.7 Dividing Fractions

Key Terms

1. reciprocal
2. indicator words
3. quotient

Objective 1

1. $\frac{4}{3}$
3. 3
5. $\frac{1}{10}$

Objective 2

7. $2\frac{2}{15}$
9. $\frac{1}{4}$
11. 6
13. $\frac{11}{15}$
15. $\frac{16}{25}$
17. $\frac{2}{9}$
19. $1\frac{53}{75}$

Objective 3

21. 54 dresses
23. 24 Brownies
25. 32 guests
27. 24 patties
29. 16 tumblers

2.8 Multiplying and Dividing Mixed Numbers

Key Terms

1. simplify
2. round
3. mixed number

Objective 1

1. 15; $13\frac{1}{3}$
3. 4; 5
5. 42; $40\frac{3}{8}$
7. 45; 51
9. 2; $2\frac{1}{2}$

Objective 2

11. $1\frac{1}{5}$; $1\frac{1}{9}$
13. 1; $1\frac{1}{4}$
15. $1\frac{3}{4}$; $1\frac{2}{3}$
17. 8; $11\frac{1}{4}$
19. $1\frac{4}{5}$; $1\frac{3}{4}$

Objective 3

21. 60 yards; 65 yards
23. 246 yards; 248 yards
25. 58 ounces; $50\frac{2}{5}$ ounces
27. $17\frac{1}{2}$ dresses; 16 dresses
29. 0 pounds; $\frac{7}{8}$ pound

Chapter 3 ADDING AND SUBTRACTING FRACTIONS

3.1 Adding and Subtracting Like Fractions

Key Terms

1. unlike fractions
2. like fractions

Objective 1

1. unlike
3. like
5. unlike

Objective 2

7. $\frac{3}{4}$
9. $1\frac{1}{2}$
11. $1\frac{1}{8}$
13. $1\frac{2}{5}$
15. $1\frac{1}{3}$
17. $\frac{7}{11}$ of the debt

Objective 3

19. $\frac{1}{5}$
21. $\frac{2}{3}$
23. $\frac{5}{14}$
25. $\frac{1}{2}$
27. $\frac{3}{7}$
29. $\frac{1}{3}$ of the garden

3.2 Least Common Multiples

Key Terms

1. least common multiple
2. LCM

Objective 1

1. 14
3. 84
5. 105

Objective 2

7. 80
9. 70

Objective 3

11. 336
13. 480
15. 420

Objective 4

17. 110
19. 108
21. 180

Objective 4

23. 18
25. 60
27. 105
29. 54

3.3 Adding and Subtracting Unlike Fractions

Key Terms

1. least common denominator
2. LCD

Objective 1

1. $\frac{5}{6}$
3. $\frac{23}{30}$
5. $\frac{43}{60}$
7. $\frac{7}{8}$
9. $\frac{19}{24}$ ton
10. $1\frac{19}{24}$ feet

Objective 2

11. $\frac{11}{15}$
13. $\frac{5}{6}$
15. $\frac{13}{21}$
17. $\frac{3}{4}$
19. $\frac{29}{54}$

Objective 3

21. $\frac{3}{8}$
23. $\frac{5}{36}$
25. $\frac{5}{8}$
27. $\frac{37}{50}$
29. $\frac{11}{24}$ of the goal

3.4 Adding and Subtracting Mixed Numbers

Key Terms

1. regrouping when subtracting fractions

2. regrouping when adding fractions

Objective 1

1. $10;\ 9\frac{4}{7}$

3. $36;\ 35\frac{17}{24}$

5. $14;\ 13\frac{32}{63}$

7. 4 cans; $4\frac{5}{24}$ cans

9. 7 hours; $6\frac{3}{8}$ hours

Objective 2

11. $4;\ 4\frac{1}{2}$

13. $0;\ \frac{35}{48}$

15. $22;\ 22\frac{1}{4}$

17. $165;\ 164\frac{11}{24}$

19. 10 hours; $8\frac{7}{8}$ hours

21. $6\text{ yd}^3;\ 5\frac{1}{24}\text{ yd}^3$

23. $14;\ 15\frac{1}{20}$

Objective 3

25. $8;\ 8\frac{1}{6}$

27. $5;\ 5\frac{7}{24}$

29. $2;\ 1\frac{5}{6}$

3.5 Order Relations and the Order of Operations

Key Terms

1. <
2. >

Objective 1

1. <
3. >
5. >
7. >
9. >

Objective 2

11. $\frac{1}{4}$
13. $4\frac{17}{27}$
15. $\frac{64}{121}$
17. $2\frac{46}{49}$

Objective 3

19. $2\frac{2}{3}$
21. $\frac{4}{25}$
23. $1\frac{5}{12}$
25. $\frac{1}{8}$
27. $1\frac{1}{6}$
29. $\frac{1}{4}$

Chapter 4 DECIMALS

4.1 Reading and Writing Decimals

Key Terms

1. decimals
2. place value
3. decimal point

Objective 1

1. $\frac{8}{10}$; 0.8; eight tenths

3. $\frac{58}{100}$; 0.58; fifty-eight hundredths

Objective 2

5. 2; 5
7. 3; 6
9. tens; ones; tenths; hundredths; thousandths

Objective 3

11. seven thousandths
13. three and fourteen ten-thousandths
15. ten and eight hundred thirty five thousandths
17. ninety seven and eight thousandths
19. 11.009
21. 300.0023
23. seven and two hundred two thousandths

Objective 4

25. $\frac{1}{1000}$
27. $20\frac{1}{2000}$
29. $\frac{19}{20}$

4.2 Rounding Decimals

Key Terms

1. decimal places
2. rounding

Objective 1

1. up

Objective 2

3. 17.9
5. 785.498
7. 54.40
9. 989.990
11. 283.05; 283.0
13. 21.77; 21.8
15. 1.44; 1.4
17. 78.70; 78.7

Objective 3

19. \$79
21. \$226
23. \$11,840
25. \$1.25
27. \$112.01
29. \$1028.67

4.3 Adding and Subtracting Decimals

Key Terms

1. front end rounding

2. estimating

Objective 1

1. 92.49	3. 105.43	5. 72.453
7. 48.35	9. 123.6802 in.	

Objective 2

11. 115.8	13. 42.566	15. 58.32
17. 24.016 ft		

Objective 3

19. 78; 82.91	21. 5; 4.838	23. 6; 5.53
25. 20 hr; 17.85 hr	27. $10; $8.71	29. 80,160 mi; 80,611.3 mi

4.4 Multiplying Decimals

Key Terms

1. factor

2. decimal places

3. product

Objective 1

1.	0.2279	3.	90.71	5.	1.5548
7.	0.0037	9.	$163.08	11.	$9.44
13.	43.0	15.	0.430	17.	0.00430
19.	348.04 sq ft	21.	$2105.99		

Objective 2

23.	300; 288.26	25.	240; 218.4756
27.	80; 43.548	29.	3200; 3033.306

4.5 Dividing Decimals

Key Terms

1. dividend
2. repeating decimal
3. quotient
4. divisor

Objective 1

1. 1.794
3. 2.359
5. 16.589

Objective 2

7. 3.796
9. 53,950.943
11. 33.3 miles per gallon

Objective 3

13. reasonable
15. unreasonable
17. unreasonable
19. reasonable

Objective 4

21. 20.31
23. 54.02
25. 96.61
27. 7.91
29. 53.548

4.6 Writing Fractions as Decimals

Key Terms

1. equivalent
2. numerator
3. denominator
4. mixed number

Objective 1

1. 6.5
3. 2.667
5. 0.091
7. 0.6
9. 4.111
11. 0.15
13. 19.708

Objective 2

15. $<$
17. $>$
19. $>$
21. $>$
23. 0.466, $\frac{7}{15}$, $\frac{9}{19}$
25. $\frac{3}{11}$, 0.29, $\frac{1}{3}$
27. $\frac{11}{13}$, 0.8462, $\frac{6}{7}$
29. 0.01666, 0.1666, 0.16666, $\frac{1}{6}$

Chapter 5 RATIO AND PROPORTION

5.1 Ratios

Key Terms

1. ratio
2. numerator; denominator

Objective 1

1. $\frac{3}{4}$ 3. $\frac{25}{19}$ 5. $\frac{17}{27}$ 7. $\frac{7}{3}$

Objective 2

9. $\frac{13}{4}$ 11. $\frac{6}{5}$ 13. $\frac{5}{6}$ 15. $\frac{3}{4}$ 17. $\frac{2}{11}$

19. $\frac{8}{5}$

Objective 3

21. $\frac{2}{7}$ 23. $\frac{9}{5}$ 25. $\frac{5}{6}$ 27. $\frac{5}{8}$ 29. $\frac{8}{1}$

5.2 Rates

Key Terms

1. unit rate
2. cost per unit
3. rate

Objective 1

1. $\frac{3 \text{ miles}}{1 \text{ minute}}$
3. $\frac{7 \text{ dresses}}{1 \text{ woman}}$
5. $\frac{15 \text{ gallons}}{1 \text{ hour}}$
7. $\frac{7 \text{ pills}}{1 \text{ patient}}$
9. $\frac{32 \text{ pages}}{1 \text{ chapter}}$

Objective 2

11. $15/hour
13. $110/day
15. $13.64/hour
17. $\frac{1}{2}$ crate/minute; 2 minutes/crate
19. $9.18/hour
21. $2.58/share
23. approximately 62 miles/hour

Objective 3

25. 16 ounces for $0.89
27. 10 for $4.19
29. 5 cans for $2.75

5.3 Proportions

Key Terms

1. proportion
2. cross products

Objective 1

1. $\frac{11}{15} = \frac{22}{30}$
3. $\frac{24}{30} = \frac{8}{10}$
5. $\frac{14}{21} = \frac{10}{15}$
7. $\frac{1\frac{1}{2}}{4} = \frac{21}{56}$
9. $\frac{6\frac{2}{5}}{12} = \frac{8}{3}$

Objective 2

11. $\frac{4}{3} = \frac{3}{4}$; false
13. $\frac{6}{5} = \frac{6}{5}$; true
15. $\frac{5}{3} = \frac{3}{4}$; false
17. $\frac{9}{5} = \frac{9}{5}$; true
19. $\frac{7}{2} = \frac{4}{1}$; false

Objective 3

21. $270 = 270$; true
23. $396 = 264$; false
25. $165\frac{3}{5} = 165\frac{3}{5}$; true
27. $114 = 342$; false
29. $12.814 = 12.07$; false

5.4 Solving Proportions

Key Terms

1.	proportion	2.	cross products	3.	ratio

Objective 1

1.	9	3.	36	5.	21
7.	5	9.	45	11.	77
13.	18	15.	33		

Objective 2

17.	1	19.	$2\frac{1}{2}$	21.	$3\frac{3}{8}$
23.	$1\frac{1}{2}$	25.	3	27.	8
29.	6				

5.5 Solving Application Problems with Proportions

Key Terms

1. ratio
2. rate

Objective 1

1. \$112.50
3. \$15
5. 8 pounds
7. \$818.40
9. 960 miles
11. \$440
13. \$4399.50
15. \$122.50
17. \$22.50
19. \$2160
21. 480 minutes or 8 hours
23. $62\frac{1}{2}$ minutes
25. 76.8 feet
27. \$2.56
29. approximately 11 days

Chapter 6 PERCENTS

6.1 Basics of Percent

Key Terms

1. ratio
2. percent
3. decimals

Objective 1

1. 43%
3. 45%

Objective 2

5. 0.42
7. 0.04
9. 0.025
11. 0.00256

Objective 3

13. 20%
15. 56.4%
17. 550%

Objective 4

19. $19
21. $228
23. $1040

Objective 4

25. 125 signs
27. 24 copies
29. 4 homes

6.2 Percents and Fractions

Key Terms

1. lowest terms
2. percent

Objective 1

1. $\frac{3}{25}$
3. $\frac{5}{8}$
5. $\frac{1}{6}$
7. $\frac{1}{200}$
9. $1\frac{2}{5}$

Objective 2

11. 70%
13. 48%
15. 94%
17. 380%
19. 740%

Objective 3

21. 0.5; 50%
23. 0.25; 25%
25. $\frac{7}{8}$; 0.875
27. $\frac{1}{3}$; 0.333
29. $\frac{13}{40}$; 32.5%

6.3 Using the Percent Proportion and Identifying the Components in a Percent Problem

Key Terms

1. whole
2. part
3. percent proportion

Objective 1

1. $\frac{\text{part}}{\text{whole}} = \frac{\text{percent}}{100}$

Objective 2

3. 800
5. 12
7. 87.5
9. 5%

Objective 3

11. 83%
13. 42%
15. 17%

Objective 4

17. 384
19. 78
21. unknown

Objective 5

23. 29.81
25. unknown
27. unknown; $\frac{x}{1500} = \frac{7}{100}$
29. unknown; $\frac{x}{40} = \frac{15}{100}$

6.4 Using Proportions to Solve Percent Problems

Key Terms

1. cross products 2. percent proportion

Objective 1

1. 280 3. 87.5 5. 24.5

7. $84 9. $210

Objective 2

11. 300 13. 500 15. 3500

17. 800 applications 19. 150 students

Objective 3

21. 0.05% 23. 5000% 25. 55%

27. 22% 29. 85%

6.5 Using the Percent Equation

Key Terms

1. percent
2. percent equation

Objective 1

1. 644
3. 106.4
5. 1.4
7. 14 clients
9. $100.50

Objective 2

11. 160
13. 2160
15. 22.8
17. 640 gallons
19. 500 employees

Objective 3

21. 20%
23. 5%
25. 244.4%
27. 25%
29. 23.3%

6.6 Solving Application Problems with Percent

Key Terms

1. commission
2. percent of increase or decrease
3. sales tax
4. discount

Objective 1

1. \$3.50; \$53.50
3. \$6.03; \$73.03
5. 5%
7. \$810

Objective 2

9. \$155.63
11. \$3000
13. 4%
15. \$2196.72

Objective 3

17. \$30; \$170
19. \$10.28; \$195.22
21. \$1408.64
23. \$36.18

Objective 4

25. 25%
27. 131.25%
29. 11.4%

6.7 Simple Interest

Key Terms

1. rate of interest
2. interest
3. interest formula
4. simple interest
5. principal

Objective 1

1. $24
3. $2122
5. $32.40
7. $8
9. $36.90
11. $780
13. $1170
15. $120
17. $63.75

Objective 2

19. $3075
21. $1125
23. $27,720
25. $32,548
27. $2149
29. $1230

6.8 Compound Interest

Key Terms

1. compound interest 2. compound amount 3. compounding

Objective 1; Objective 2

1. $2100
3. $2315.25; $315.25

Objective 3

5. $4867.20
7. $4920
9. $3595.52

Objective 4

11. $1262.50
13. $11,277
15. $60.47
17. $19,815.73

Objective 5

19. $1272.30; $272.30
21. $20,724.48; $7924.48
23. $34,730.06; $13,330.06
25. $111,212.40; $33,212.40
27. $1302.30; $302.300
29. $14,282.10; $5282.10

Chapter 7 MEASUREMENT

7.1 Problem Solving with U.S. Customary Measurements

Key Terms

1. metric system
2. U.S. customary measurement units

Objective 1

1. 12
3. 2
5. 5280
7. 3

Objective 2

9. 4 yards
11. 210 inches
12. 56 ounces
13. 10,000 pounds
15. 1.25 or $1\frac{1}{4}$ minutes
17. 0.83 or $\frac{5}{6}$ yard

Objective 3

19. 3.5 or $3\frac{1}{2}$ gallons
21. 3.75 or $3\frac{3}{4}$ pounds
23. 21,120 feet
25. 120 hours

Objective 4

27. $5.31 per pound
29. $6.36 per pound

7.2 The Metric System—Length

Key Terms

1. prefix
2. metric conversion line
3. meter

Objective 1

1. mm
3. m
5. m
7. mm
9. km

Objective 2

11. 0.07 m
13. 2300 mm
15. 0.45 km
17. 60.2 cm
19. more; 3 cm

Objective 3

21. 0.636 m
23. 14.5 km
25. 0.000035 cm
27. 610 m
29. more; 201 cm

7.3 The Metric System—Capacity and Weight (Mass)

Key Terms

1.	gram	2.	liter

Objective 1

1.	mL	3.	L

Objective 2

5.	0.007 kL	7.	2500 mL	9.	836,000 L
11.	7.863 L				

Objective 3

13.	mg	15.	g

Objective 4

17.	9 kg	19.	6300 g	21.	4700 mg
23.	0.008745 kg				

Objective 5

25.	mg	27.	km	29.	kg

7.4 Problem Solving with Metric Measurement

Key Terms

1. gram 2. meter 3. liter

Objective 1

1. 60 servings
3. $12.99
5. 500 bottles
7. 8.33 m
9. 2.45 L
11. 3 pills
13. 5600 mL
15. 1250 g
17. 840 g
19. 20 units
21. above the speed limit; 6 km per hour
23. $18.57
25. the $28 case that holds sixteen 600 mL bottles; $2.92
27. 188.5 km
29. yes; 7 L

7.5 Metric–U.S. Customary Conversions and Temperature

Key Terms

1. Celsius
2. Fahrenheit

Objective 1

1. 21.3 lb
3. 4.8 m
5. 10.2 kg
7. 40.6 cm
9. 2.6 qt
11. $11.44
13. gallon bottle at $3.60

Objective 2

15. 13ºC
17. 65ºC
19. 300ºF

Objective 3

21. 17ºC
23. 27ºC
25. 86ºF
27. 842ºF
29. 204ºC

Chapter 8 GEOMETRY

8.1 Basic Geometric Terms

Key Terms

1. ray
2. perpendicular lines
3. obtuse angle
4. point
5. angle
6. line
7. acute angle
8. degrees
9. straight angle
10. parallel
11. line segment
12. intersecting lines
13. right angle

Objective 1

1. line segment, $\overline{CD}$
3. line, $\overleftrightarrow{EF}$
5. ray, $\overrightarrow{CB}$
7. ray, $\overrightarrow{RS}$

Objective 2

9. parallel
11. parallel

Objective 3

13. $\angle HEI$
15. $\angle NKL$
17. $\angle WSV$

Objective 4

19. right
21. acute
23. straight
25. right

Objective 5

27. parallel
29. perpendicular

8.2 Angles and Their Relationships

Key Terms

1. vertical angles
2. congruent angles
3. supplementary angles
4. complementary angles

Objective 1

1. 78°
3. 18°
5. 86°
7. 164°
9. 142°
11. $\angle BAC$ and $\angle CAD$; $\angle FAE$ and $\angle EAD$
13. $\angle LKM$ and $\angle MKN$; $\angle MKN$ and $\angle NKO$; $\angle NKO$ and $\angle OKL$; $\angle OKL$ and $\angle LKM$

Objective 2

15. $\angle NOM \cong \angle POQ$; $\angle MOQ \cong \angle PON$
17. $\angle BLV \cong \angle WLC$; $\angle CLV \cong \angle WLB$
19. $\angle CAD \cong \angle BAE$
21. 33°
23. 42°
25. 50°
27. 37°
29. 73°

8.3 Rectangles and Squares

Key Terms

1. area
2. rectangle
3. perimeter
4. square

Objective 1

1. $P = 24$ cm; $A = 32$ cm^2

3. $P = 36$; $A = 17$ cm^2

5. $P = 22$ yd; $A = 29\frac{1}{4}$ yd^2

7. $P = 281.2$ cm; $A = 3859.68$ cm^2

9. $A = 7248$ ft^2

Objective 2

11. $P = 36$ m; $A = 81$ m^2

13. $P = 31.2$ ft; $A = 60.84$ ft^2

15. $P = 5\frac{3}{5}$ in.; $A = 1\frac{24}{25}$ in.^{2}

17. $P = 12.4$ cm; $A = 9.61$ cm^2

19. $P = 18\frac{2}{3}$ mi; $A = 21\frac{7}{9}$ mi^2

Objective 3

21. $P = 30$ ft; $A = 18$ ft^2

23. $P = 42$ yd; $A = 50$ yd^2

25. $P = 24$ cm; $A = 31$ cm^2

27. $P = 42$ yd; $A = 54$ yd^2

29. $P = 56$ cm; $A = 171$ cm^2

8.4 Parallelograms and Trapezoids

Key Terms

1. parallelogram
2. trapezoid
3. perimeter
4. area

Objective 1

1. 168 m
3. 30 ft
5. 22 cm
7. 27 in.^2
9. 713 yd^2
11. $11\frac{1}{4}$ m^2
13. 310 ft^2
15. $780

Objective 2

17. 708.8 cm
19. 1106 m^2
21. 60 in.^2
23. $9\frac{5}{8}$ in.^2
25. 4190 cm^2
27. 15,504 ft^2
29. $700

8.5 Triangles

Key Terms

1.	triangle	2.	base	3.	height

Objective 1

1.	25 yd	3.	$24\frac{1}{2}$ ft	5.	37.2 ft
7.	$10\frac{3}{4}$ ft	9.	17.7 m		

Objective 2

11.	1260 m^2	13.	$21\frac{3}{4}$ ft^2	15.	15.81 m^2
17.	510 m^2	19.	534 m^2	21.	1940 yd^2

Objective 3

23.	17°	25.	81°
27.	51°	29.	45°

8.6 Circles

Key Terms

1. radius
2. circumference
3. circle
4. π (pi)
5. diameter

Objective 1

1. 86 m
3. 13.25 m
5. 4 ft
7. $6\frac{1}{4}$ yd

Objective 2

9. 94.2 m
11. 28.3 yd
13. 188.4 cm

Objective 3

15. 43.0 m^2
17. 22.3 yd^2
19. 57 cm^2
21. 2101.3 m^2
23. 18.2 ft^2
25. $320.87

Objective 4 *Other answers are possible.*

27. polynomial; polyglot
29. octagon; octopuse

8.7 Volume

Key Terms

1. rectangular solid
2. cylinder
3. sphere
4. volume
5. pyramid
6. cone

Objective 1

1. 2744 in.^3
3. 176 in.^3
5. 48 m^3
7. 2310 m^3

Objective 2

9. 56.5 ft^3
11. 0.9 in.^3
13. 3267.5 ft^3

Objective 3

15. 471 ft^3
17. 0.1 km^3
19. 678.2 cm^3
21. 1433.5 in.^3

Objective 4

23. 121.3 m^3
25. 333.3 m^3
27. 25.8 m^3
29. $22{,}344 \text{ m}^3$

8.8 Pythagorean Theorem

Key Terms

1.	right triangle	2.	hypotenuse	3.	legs

Objective 1

1.	4.123	3.	1.414
5.	8.660	7.	12.042

Objective 2

9.	9.8 in.	11.	12.1 in.	13.	1.0 in.
15.	8.5 in.	17.	13 ft		

Objective 3

19.	9.5 ft	21.	5 ft	23.	30 ft
25.	8 ft	27.	7.2 ft	29.	19.1 ft

8.9 Similar Triangles

Key Terms

1. congruent
2. similar triangles

Objective 1

1. $\overline{AB}$ and $\overline{PQ}$; $\overline{AC}$ and $\overline{PR}$; $\overline{BC}$ and $\overline{QR}$;
   $\angle A$ and $\angle P$; $\angle B$ and $\angle Q$; $\angle C$ and $\angle R$

3. $\overline{PN}$ and $\overline{SR}$; $\overline{NM}$ and $\overline{RQ}$; $\overline{MP}$ and $\overline{QS}$;
   $\angle P$ and $\angle S$; $\angle N$ and $\angle R$; $\angle M$ and $\angle Q$

5. $\overline{HK}$ and $\overline{RS}$; $\overline{GH}$ and $\overline{TR}$; $\overline{GK}$ and $\overline{TS}$;
   $\angle H$ and $\angle R$; $\angle G$ and $\angle T$; $\angle K$ and $\angle S$

7. $\frac{3}{2}, \frac{3}{2}, \frac{3}{2}$

Objective 2

9. $a = 6\frac{2}{3}; b = 20$
11. $a = 15; b = 9$
13. $x = 12.75; y = 15$
15. 24
17. 54 in.; 67.5 in.
19. 24; 32

Objective 3

21. 39 ft
23. 36 m
25. 7.5 ft
27. 132 yd
29. 10 m

Chapter 9 BASIC ALGEBRA

9.1 Signed Numbers

Key Terms

1. signed numbers
2. opposite of a number
3. absolute value
4. negative numbers

Objective 1

1. +17
3. +830
5. −120

Objective 2

7. 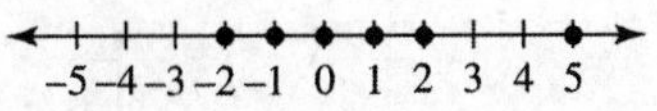 −5 −4 −3 −2 −1 0 1 2 3 4 5

9. 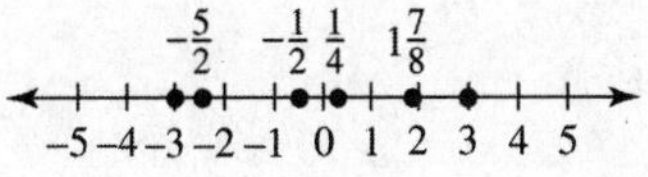 $-\frac{5}{2}$ $-\frac{1}{2}$ $\frac{1}{4}$ $1\frac{7}{8}$
 −5 −4 −3 −2 −1 0 1 2 3 4 5

11. 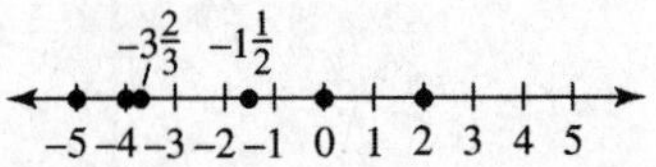 $-3\frac{2}{3}$ $-1\frac{1}{2}$
 −5 −4 −3 −2 −1 0 1 2 3 4 5

Objective 3

13. $<$
15. $>$
17. $>$

Objective 4

19. 11
21. −8.23
23. $\frac{5}{7}$

Objective 5

25. −3
27. 3
29. $\frac{2}{3}$

9.2 Adding and Subtracting Signed Numbers

Key Terms

1. absolute value
2. additive inverse

Objective 1

1. 15
3. -8

Objective 2

5. 7
7. -2.82
9. $\frac{3}{4}$
11. $-3\frac{1}{4}$

Objective 3

13. 0
15. -281
17. -2.7

Objective 4

19. -1.2
21. $\frac{5}{4}$ or $1\frac{1}{4}$
23. 13.2

Objective 5

25. -11.5
27. $-\frac{1}{3}$
29. 3.1

9.3 Multiplying and Dividing Signed Numbers

Key Terms

1. factors
2. quotient
3. product

Objective 1

1. -24
3. -96
5. -32.2
7. -29.76
9. -4
11. $-\frac{5}{4}$ or $-1\frac{1}{4}$
13. -8
15. -5.92

Objective 2

17. 90
19. $\frac{12}{5}$ or $2\frac{2}{5}$
21. $\frac{9}{2}$ or $4\frac{1}{2}$
23. $\frac{9}{7}$ or $1\frac{2}{7}$
25. $\frac{1}{2}$
27. 4.55
29. $\frac{1}{16}$

9.4 Order of Operations

Key Terms

1. order of operations
2. base
3. exponent

Objective 1

1. –20
3. 2
5. 30
7. –3
9. 14

Objective 2

11. –3
13. 45
15. 31
17. –6
19. –18

Objective 3

21. 9
23. $\frac{16}{21}$
25. –1
27. –1
29. 2

9.5 Evaluating Expressions and Formulas

Key Terms

1. expression
2. variable

Objective 1

1. expression
3. expression

Objective 2

5. −17
7. −15
9. −12
11. −4
13. 1
15. −3
17. $\frac{12}{16}$ or $\frac{3}{4}$
19. 20
21. 36
23. 195
25. 113.04
27. 33.49
29. −5

9.6 Solving Equations

Key Terms

1. equation
2. multiplication property of equations
3. solution
4. addition property of equations

Objective 1

1. not a solution
3. not a solution
5. solution
7. solution
9. not a solution

Objective 2

11. $p = 14$
13. $m = 1$
15. $x = \frac{2}{3}$
17. $k = 5\frac{1}{3}$
19. $x = 5.61$

Objective 3

21. $k = 32$
23. $r = -36$
25. $m = -1.1$
27. $m = -32$
29. $k = 2.1$

9.7 Solving Equations with Several Steps

Key Terms

1. like terms
2. distributive property

Objective 1

1. $p = 1$
3. $y = 11$
5. $x = -2$
7. $z = 6$

Objective 2

9. $-10 - 5a$
11. $5 - x$
13. $-32 + 4x$

Objective 3

15. $2m$
17. $-5a$
19. $2.3a$

Objective 4

21. $y = 2$
23. $a = 0$
25. $y = 1$
27. $z = 3$
29. $d = \frac{28}{12} = \frac{7}{3}$ or $2\frac{1}{3}$

9.8 Using Equations to Solve Application Problems

Key Terms

1. indicator words
2. sum; increased by
3. product; double
4. quotient; per
5. difference; less than

Objective 1

1. $9 + x$
3. $3x$
5. $3x + 3$
7. $10x - 6x$

Objective 2

9. $45 - 2x = 35;\quad x = 5$
11. $2x - 3 = -17;\quad x = -7$
13. $9x - 7x = 16;\quad x = 8$
15. $\frac{1}{2}x + 4 = 10;\quad x = 12$

Objective 3

17. −1
19. 8 cm
21. Jerry: 36 years old; Marie: 12 years old
23. −3
25. 38 cm
27. 250 miles
29. 21 inches

Chapter 10 STATISTICS

10.1 Circle Graphs

Key Terms

1. circle graph
2. protractor

Objective 1

1. \$10,400
3. \$400
5. $\frac{300}{1400}$ or $\frac{3}{14}$

Objective 2

7. business
9. $\frac{1800}{11,600}$ or $\frac{9}{58}$
11. $\frac{600}{2400}$ or $\frac{1}{4}$
13. \$285,000
15. \$95,000
17. \$142,500
19. $\frac{142,500}{95,000}$ or $\frac{3}{2}$
21. $\frac{285,000}{237,500} = \frac{6}{5}$
23. computer science
25. $\frac{18\%}{6\%}$ or $\frac{576}{192}$ or $\frac{3}{1}$
27. 29%

Objective 3

29. (a) mysteries: 108°; biographies: 54°; cookbooks: 36°; romance novels: 90°; science: 54°; business: 18°

(b)

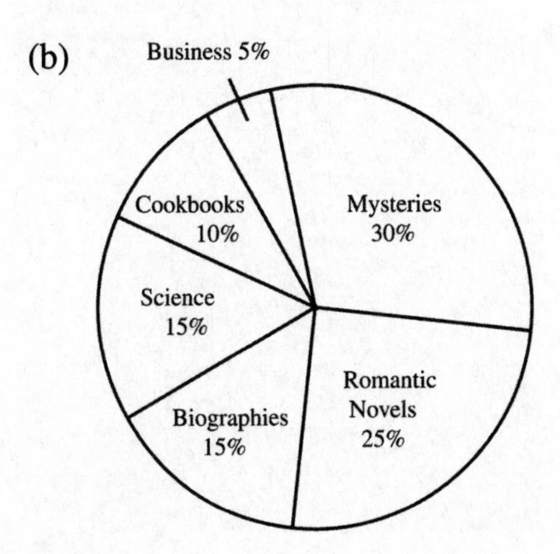

10.2 STATISTICS

Key Terms

1. line graph
2. double-bar graph
3. bar graph
4. comparison line graph

Objective 1

1. 1000 students
3. 1800 students
5. 2006
7. 400 students

Objective 2

9. 350 female freshman
11. $\frac{500}{350}$ or $\frac{10}{7}$
13. $\frac{850}{550}$ or $\frac{17}{11}$

Objective 3

15. September
17. $20
19. $\frac{60}{20}$ or $\frac{3}{1}$
21. $30

Objective 4

23. $1,000,000
25. $2,500,000
27. $3,000,000
29. 2004

10.3 Frequency Distributions and Histograms

Key Terms

1. histogram
2. frequency distribution

Objective 1

1. || ; 2
3. |||| | ; 6
5. |||| ; 4

Objective 2

7. |||| ; 5
9. ||| ; 3
11. ||| ; 3
13. 110–129
15. | ; 1
17. |||| || ; 7
19. | ; 1
21. 80–89 and 120–129
23.

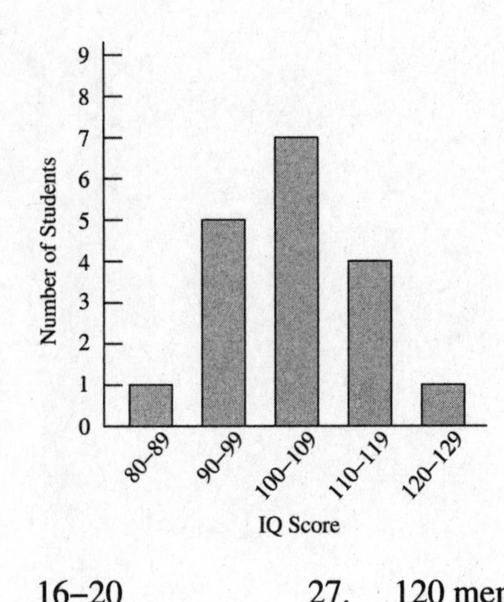

25. 16–20
27. 120 members
29. 30 members

10.4 Mean, Median, and Mode

Key Terms

1. dispersion	2. mode	3. weighted mean
4. mean	5. median	6. bimodal
7. range		

Objective 1

1. 60.3	3. 64.5
5. 52.7	7. 5.4

Objective 2

9. 18.6	11. 4.7
13. 3.1	15. 2.5

Objective 3

17. 232	19. 632
21. 25	23. 239.5

Objective 4

25. 3	27. 24, 35, 39	29. 8